Courtesy of the author

ABOUT THE AUTHOR

Dr. Graham Tattersall is a freelance engineer working on projects as diverse as automated image recognition, fault analysis systems for trains, and enhancement of ultrasound images.

GEEKSPEAK

GEEKSPEAK

A Guide to Answering the Unanswerable, Making Sense of the Insensible, and Solving the Unsolvable

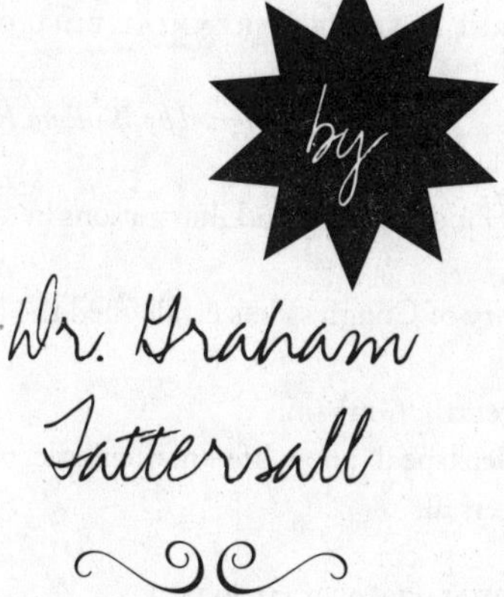

Dr. Graham Tattersall

NEW YORK • LONDON • TORONTO • SYDNEY • NEW DELHI • AUCKLAND

HARPER PERENNIAL

First published in Great Britain in 2007 by Fourth Estate.

A hardcover edition of this book was published in 2008 by Collins, an imprint of HarperCollins Publishers.

FIRST HARPER PERENNIAL EDITION PUBLISHED 2010.

Designed by William Ruoto

Hand lettering and illustrations by Geoff Westby 2007

The Library of Congress has catalogued the hardcover edition as follows:

Tattersall, Graham.
Geekspeak : how life+mathematics=happiness / Graham Tattersall.
p. cm.
ISBN 978-0-06-162924-2
1. Mathematics—Popular works. I. Title.
QA93. T38 2008
510—dc22 2008016134

ISBN 978-0-06-162678-4 (pbk.)

10 11 12 13 14 WBC/RRD 10 9 8 7 6 5 4 3 2 1

This book is written in memory of my father, who helped raise six children as well as quietly going through life inventing, analyzing, and mending anything and everything.

Contents

Photograph Acknowledgments

All photos, unless stated, taken from Stock.XCHNG.com

Introduction: Living with Numbers: #347053, Aleksander Milosevic, Belgrade, Serbia

1. Scrabbling for Words: #650192, Steve Woods/Pinpoint, Essex, UK
2. Pumping Iron: #313419, Josef Lluis Caldentey, Spain
3. Safe as Houses: Author
4. Well Connected: #368906, Cory McKenzie, Canada
5. Fatal Attraction: #696930, Marcos Santos, Brazil
6. Home Alone: #656412, Liv Crazy, Australia
7. Beam Me Up, Scotty: #757924, Artiom, Ukraine (StockExpert)
8. Hidden Death: #97469, Ken Kiser, US
9. It's for You: #6926, Stephen Gibson/BudgetStockPhoto .com, Australia
10. War Chest: #522105, Marcin Barlowski, Poland
11. When the Wind Blows: #739813, George Bosela, US
12. You're Rubbish: #315128, Liton Ali, London, UK
13. The Cunning Fox: #769992, Lynne Lancaster (weirdvis), UK

14. Fly Wheels: #735390, Asif Akbar, India
15. Bus Stop: #198141, Tomasz Kowalczyk, Poland
16. Stirring Airs: #442256, Steve Ford Elliot, Eire
17. Dream Flight: #748302, Martin Brooks, UK
18. Processing Power: #370098, Max Brown, Sydney, Australia
19. Soul Mates: #264203, Paul Preacher, London, UK
20. Idiot Calculus: #282161, Anka Draganski, London, UK
21. The Ghostly Present: #708887, Robert Rosmond, New Orleans, US
22. Bad Breath: #398156, Tudou Mao, Shijiazhuang, China
23. The Final Judgment: #19734, alxm (StockExpert)
24. Heavenly Body: #707105, Wojtek Wozniak, Zabrze, Poland
25. Passing Water: #633517, Emin Ozkan, Zmin, Turkey
26. The Man in White: #700514, Joan Koele, Netherlands

Introduction

LIVING WITH NUMBERS

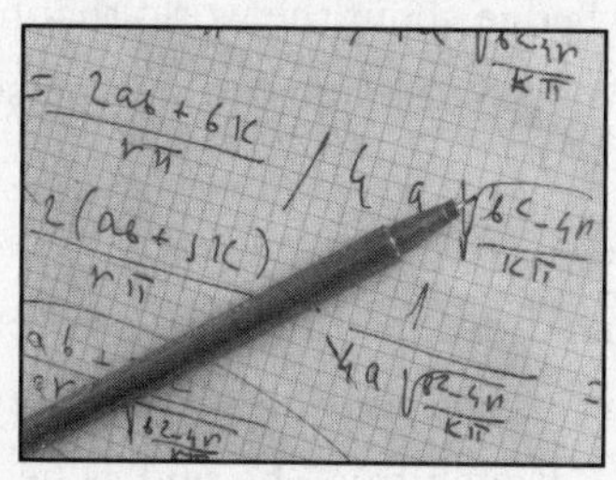

How much can you work out about your own world?

For as long as I can remember, I've been fascinated by how things *work*. There's a whole world of cogs, circuit boards, and equations behind familiar objects and events, or the figures and statistics in the media. I can remember the exact moment when this fascination was born: it was as an eight-year-old child on a family trip to the Scottish Highlands. Frantic to distract his squabbling children, my father announced that "The first person to tell me the weight of that mountain gets to sit in the front of the car." I took on the task, excitedly shouting out wrong answers.

For me, this was power—working out a new number, a new piece of knowledge about the world that hadn't existed ten seconds earlier. And, it all happened in a person's head. It was amazing.

And that was it. I became a Geek.

A family friend dropped by after our return. He was an amateur radio enthusiast, and his car was festooned with all sorts of antennas. "Watch this," he said as he placed his finger at the bottom end of an antenna.

The faint, sweet smell of burning flesh drifted toward me as the current flowed through his skin, heating and burning its outer surface but without causing any pain or real damage.

I was transfixed. How had that happened? Why hadn't it hurt? How deep had the current penetrated? And how much energy had been spent in his finger? Here were more questions, more knowledge to attain, more facts and figures, all calculable inside a person's head.

The fascination with the simple conjuror's trick of the burning skin has stayed with me, as a desire to explore the hows and whys lurking behind everyday life. And that, in a nutshell, is what this book is about: how to analyze and understand your world.

It's also about your ability to judge and check "expert" opinion for yourself rather than take it for granted—about using your numeracy to be better informed.

Have a look at a couple of expert assertions. Do you believe them? How would you judge their validity?

> Developed nations should focus on the development of wind and wave power to meet its electricity generation needs.
>
> Hormone replacement therapy (HRT) greatly increases the risk of breast cancer.

Neither statement can be tested without using numbers: the amount of power that can be generated by a wind turbine and the electricity needed to run a country; and the statistical significance and experimental controls used to make the HRT assertion.

But the numbers themselves are of little value. To understand the issues, the assumptions, caveats, and uncertainties need to be understood. And the only way to understand all those is to do the calculation yourself.

Many people think that to perform that kind of calculation you need to be an expert in a given field—which is why we tend to rely on experts. You might decide that it's best to leave it to the professionals, the politicians, and the people in white coats.

But there is another way: a path of knowledge, learning, and understanding—the way of the Geek. Not an obsessive, narrowly interested, malodorous Geek but a nice, thoughtful, sweet-smelling Geek, the kind you'd like as a friend.

A true Geek is interested in the mathematics of body size, and of God, sex, food, and a whole load of other things that also interest quite normal people.

And this is the rub. We live surrounded by figures: the power of a wind turbine, the amount the average family spends on food, the fuel consumption of a Boeing 747, the weight of sewage you create each year. Not being able to estimate such figures yourself—not speaking any Geek—is like living in a foreign country and not being able to communicate.

How can you trust statements made by academics, architects, scientists, and politicians without checking the underlying numbers?

You can't!

But you *can* work it out for yourself. And what's more, once you've done that it's better than just knowing the fact. You'll understand the basis and limits of the truth.

This book is about empowerment, by combining common sense, straightforward arithmetic, and a little questioning of received wisdom.

This book shows you how to speak Geek.

1

SCRABBLING FOR WORDS

How big is your vocabulary?

You know thousands of words. Jane Austen uses more than 6,000 different words in *Pride and Prejudice*, and you can read them all without the slightest problem. In fact, your passive reading vocabulary probably exceeds 10,000 words.

On the other hand, your active vocabulary—the words you use in everyday speech—is much more limited. On an average day you'll probably get by on a few hundred words. And those words say a lot about you: your sex, age, and social class.

In the early 1990s, recordings were made of conversations and used to build a database of words in the English language. The database, held at the University of Lancaster, contains over 100 million words spoken by men and women of all ages and occupations. Three researchers, Paul Rayson, Geoffrey Leech, and Mary Hodges pulled out some interesting facts from this data. One of the most startling is the difference between the kind of words used by men and those used by women. There are certain words that act as fingerprints, showing that a conversation is between two men, or between two women.

These are the top three fingerprint words in women's conversation by academic researchers:

she
her
said

And the three words most characteristic of man-to-man conversation:

fucking

er

the

Those top three female words are instantly recognizable as typical of "girl talk." Just eavesdrop on a conversation between two women chatting near the office coffee machine: "And *she said* that *her* friend was really upset . . . And I *said* to *her* . . ."

As for the men, here are a couple of guys leaning over the open hood of a car: "What's that *fucking* wire doing?" "*Er*, dunno. *The* battery's dead."

All joking aside, the journey from the equality of baby burbling to speaking in ways that encode your gender, age, and social status takes two or three decades, but you can go back to the first moments after your birth quite easily. Start by letting every muscle in your mouth and lips go slack. Now make a noise.

That grunt is called the *schwa*. It is the most basic neutral vowel sound, and it sounds similar, though not identical, when uttered by people with different mother tongues.

In the first few months after your birth, you'll start to babble, and by the time you're coming up to your first birthday you'll have a few words. Those words use vowel sounds such as *u* as in "mum" and *a* as in "man." They'll be bracketed by primitive consonants or nasal sounds such as *m* and *n* to create important words such as "momma" and "dadda."

Fast-forward to the age of around eighteen months, and you'll be making much more complex sounds by articulating most vowels and consonants, and introducing *l* and *n* sounds.

The extremes of the vowel sounds are the cardinal points of your language. In English, they range from *a* as in "cat" and *i* as in "hit," to *oo* as in "hoot" and *aw* as in "saw." You can utter a kind of sound circle with the cardinal vowels. Voice them in sequence and you'll find that the sound changes smoothly from one vowel to the next.

Counting vowels, consonants, nasals, and *l* and *r* sounds, a fully developed English speaker can recognize at least forty-five basic sounds. They are called *phonemes.*

It used to be thought that each phoneme was a distinct acoustic event, but it is now accepted that many are psycho-perceptual. For example, the stop consonant *pp* in the word "apple" does not exist by itself. A stop consonant is the sound we think we hear ourselves saying when we use our mouth to rapidly stop or start a sound. The *pp* in "apple" is the sound made when we quickly close our lips to stop making the *a* sound, and then explosively open them again to continue with the *le* sound.

We perceive the stop consonant as an actual sound that exists between the *a* and the *le.* But if you look at the sound wave of someone saying "apple," you'll see a period of silence in the middle of the word. That's the *pp* in "apple." It doesn't exist: it's simply perceived because of the way the *a* and *le* are stopped and started.

One of the drawbacks of growing up speaking your mother tongue is losing responsiveness to speech sounds in other languages, as some Japanese researchers demonstrated. They played sounds to infants while monitoring the frequency with which the infants sucked on a pacifer. They sucked more often when there was a recognizable stimulus such as their mother's face or a familiar sound.

Newborns, who had barely been exposed to their mother tongue, sucked rapidly when they heard any of a wide variety of sounds drawn from many languages. Older infants sucked rapidly only when they heard sounds used in their mother tongue. The researchers inferred that infants lose the ability to distinguish certain sounds when they start to learn a language in which those sounds are absent.

Now, hopefully many years after you stopped sucking because something seemed familiar, you understand, speak, and read thousands of words. It's rather strange that we bother, when just a few hundred words are sufficient for our daily lives.

So, how many words do *you* know?

It's possible to work out the size of your passive vocabulary. One approach is to go through every entry in a dictionary and check off every word you know. But if you've got other things to do, there's a way that gives a good estimate in a much shorter time: statistical sampling.

The idea behind statistical sampling is the same as used in surveys of, for example, voters. The nation's voting pattern could be found by asking all 150 million intended voters about their plans for the voting booth. More practically, a representative sample of voters is questioned perhaps just 1,000 people carefully selected to represent all the localities and social groups in the country.

The same approach can be used to estimate your vocabulary. Sample the "population" of words by opening the dictionary at random a hundred times. Each time, look at the first entry at the top of the page. Do you know the meaning of this word? If the answer is yes, add one to your score. At the end of the exercise, divide your score by the sample size of one hundred to get an estimate of the fraction of words you know. Multiply that fraction by the total number of words in the dictionary to estimate your vocabulary size.

This method works, but you need to be careful: how many times should you dip into the dictionary at random to get a good estimate? Say you do the test twice and find that you know the first word, but not the second. That means that you know 50 percent of the words in the tiny bit of the dictionary you examined.

But common sense tells you that this estimate is unreliable. It is true that you might know half the dictionary, but it is also possible that you know 10 percent or 90 percent of all the words. The two words you chanced upon might have been unusually uncommon or unusually common. Two is not a representative sample.

Do the trial 10 times, and confidence in the result is greater; 100 times, even better. If you did the trial 1,000 times and found that you knew 500 words, you could argue quite strongly that you really do know about half of all the words in the dictionary.

To estimate your vocabulary you'll need to know the total number of words in the dictionary—preferably without having to count them. This too is quite easy. Look up the number of the last page in the dictionary, and take that as the number of pages. Next, open the dictionary at random and count the number of different words listed on that page. Multiply the number of pages by the number of words per page, and you have an estimate of the number of words in the dictionary.

I thought I'd better test myself using this statistical sampling technique. The dictionary I used has about 60 entries on each page, and more than 800 pages. That's around 48,000 words altogether.

I opened the dictionary 125 times, and made a check on a piece of paper if I knew the meaning of the word at the top of the page, and an *X* if I didn't. Like me, you'll probably find it hard to stop yourself jumping ahead to other entries if the first is unfamiliar. Don't—that's cheating and invalidates the statistical sampling!

The result: there were 25 words whose meaning I didn't know. On that basis, my passive vocabulary is 48,000 multiplied by 100/125. That's around 40,000 words. It sounds high, but it includes all the possible extensions of the stem of each word. For example, take the word "abstract." The dictionary will include "abstractedly," "abstractedness," and so on. The number of stem words I know is a lot less than 40,000.

Still, I'm feeling pretty good about myself, so I'm going to exercise my gigantic male vocabulary by introducing the next chapter:

"The, er, next chapter is, er, fucking interesting . . ."

SPEAK GEEK

— A practical application —

"IT IS A TRUTH UNIVERSALLY ACKNOWLEDGED THAT A SINGLE MAN IN POSSESSION OF A GOOD FORTUNE MUST BE IN WANT OF A WIFE."

—JANE AUSTEN

Some authors are instantly recognizable from their vocabulary. For example, everyone recognizes the style of Jane Austen, and many would say that her writing's distinguishing feature is its abundance of long words. But is this true? A bit of statistical analysis can reveal the answer.

The four longest words used by Jane Austen in *Pride and Prejudice* have 16 or 17 characters. They are "superciliousness," "communicativeness," "disinterestedness," and "misrepresentation." But just looking at the longest words is not enough: we

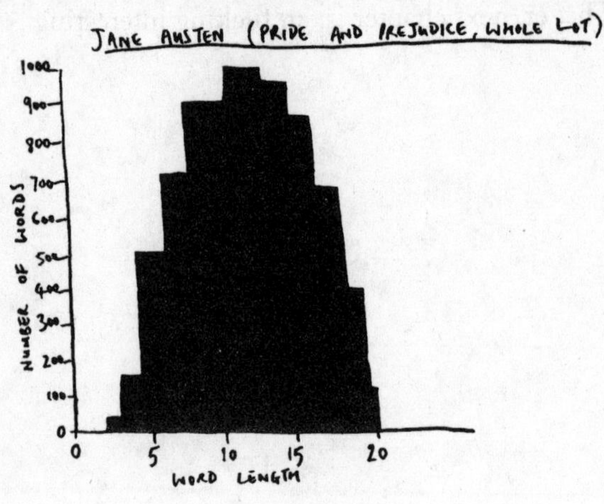

need to examine the distribution of word lengths over her entire vocabulary, as shown in the graph below:

For comparison, here is the "fingerprint" of the writer Ian McEwan, showing that his vocabulary includes many shorter words:

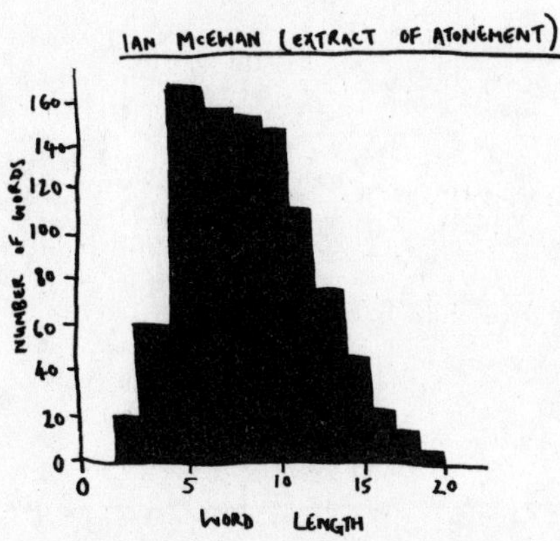

And, what about this book? In this work I intend to speak with candor, and without misrepresentation or superciliousness, of the accomplishments of the irreproachable retrospections. . . .

2

PUMPING IRON

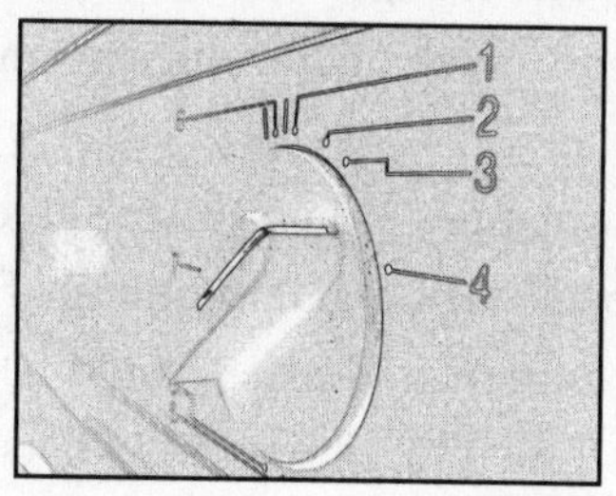

Are you as powerful as a washing machine?

For the price of a few cups of oil, men can be transformed into mechanical supermen. Next time you pass major construction or a large building site, watch the hydraulic rams on a mechanical digger pushing the bucket and scooping up one-ton heaps of sand, all at the twist of the driver's wrist. Those rams are extensions of his limbs: he is a superman.

Mechanical power is often quantified as *horsepower*, a word coined by the eighteenth-century engineer James Watt, the man whose work changed steam engines from profligate steam guzzlers into much more efficient and powerful machines.

In Watt's day, ponies or horses were used to turn a windlass that hoisted buckets of coal up a mineshaft. He would have wanted to know how many horses would be needed to lift a bucket in a given time. Watt knew that a horse could pull with a force of about 180 pounds, and that it could walk a total distance of around 180 feet each minute while pulling the load. That became his definition of horsepower: one horsepower is the power needed to move a force of 180 pounds through a distance of 180 feet every minute.

You might think of it this way: an average man weighs around 180 pounds, so with a suitable pulley and rope, a horse could hoist him 180 feet into the air in about one minute. The Eiffel Tower in Paris is 986 feet high. If we could position our hoisting pulley at the top of the Tower, our man would be dangling almost one-fifth of the way up after one minute.

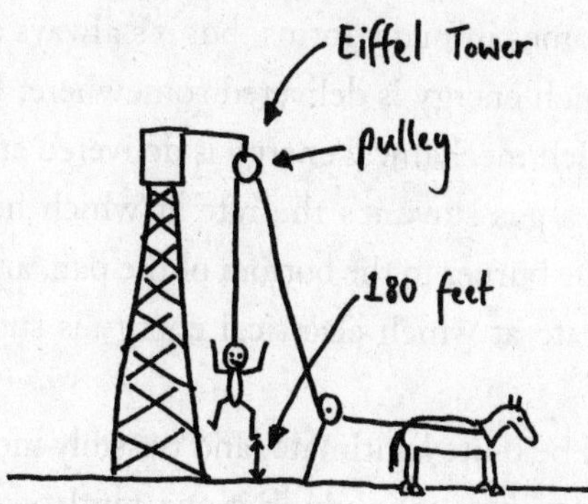

Connect the same pulley rope to the 60 hp engine in your car, and you could hoist the same man to the top of the Eiffel Tower in less than six seconds, although he might not have much stomach for the view at the top.

What about men instead of horses? A simple way of measuring your own horsepower would be to tie the pulley rope around your own waist, take the strain, and see how long it takes you to hoist the 180-pound man 180 feet.

A 180-pound man is too much for most of us to lift, so let's replace him with a small child weighing, say, a quarter of that, 45 pounds. If you managed to hoist the child through 180 feet in one minute, you would have a power of one-quarter of a horsepower.

In reality, even if someone was willing to lend you their child in the service of science, most people would have difficulty in performing the task in less than two and a half minutes. So, a man's power is nearer one-quarter divided by two and half, which is one-tenth of a horsepower.

Nowadays power is usually measured in watts rather than horsepower. We've just changed from using Watt's own term, horsepower, to using his own name as our standard unit of power. There are 746 watts in one horsepower.

Power comes in many forms, but it's always a measure of the rate at which energy is delivered somewhere. For a car, it's the rate at which mechanical energy is delivered at the engine's flywheel. For a gas stove it's the rate at which heat energy is delivered by the burner to the bottom of the pan, and for a lightbulb it's the rate at which electrical energy is supplied to the bulb.

It would be quite legitimate, and possibly more meaningful, to rate, say, a 75-watt lightbulb as one-tenth of a horsepower. A label of 1/10 hp on the bulb would indicate that one horse turning a windlass connected to an electrical generator could light ten such bulbs. More sobering, it shows that one athlete turning a windlass or a treadmill could manage to keep just one 75-watt bulb burning.

I'm guilty of sometimes having up to 300 watts-worth of lightbulbs switched on in my house during the evening. In a pre-fossil-fuel era, I would have needed four slaves walking continuously on a treadmill to keep them alight. But don't be too smug. To boil water in your 3 kW electric kettle you would need forty slaves.

Using watts as a unit of power has probably contributed to the divorce between our understanding of machine power and of human power. Labeling commonplace machines and devices in manpower instead of watts might keep us much more aware of our dependence on fossil fuel.

In a world without fossil fuel, the unit of power might be the "slave." If a slave is equal to 1/10 hp, your car has a 600-slave engine, your water heater is rated at 40 slaves, and your fridge is about one slave. Yes, there would need to be one slave pedaling a generator 24/7 to keep your food cold.

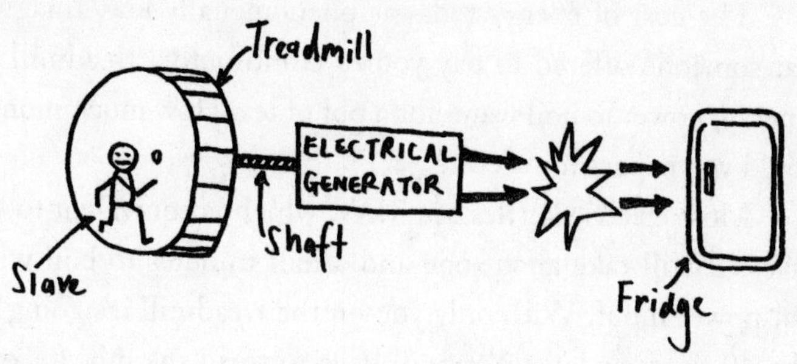

Machine	Power in watts	Power in slaves
Pocket flashlight	1	1/75
Phone charger	1.5	1/50
Portable FM radio	7.5	1/10
Digital radio	10	0.0
Low-energy lightbulb	18	1/4
TV on standby	25	1/3
Modern fridge	75 (daily average)	1
Incandescent lightbulb	75	1
TV switched on	75	1
Electric kettle	3,000	40
Small oil central-heating system	15,000	200
Your car	50,000	600
Tractor digger	150,000	2,000
High-speed train	4,500,000	60,000
747 jumbo jet	90,000,000	1,200,000

The cost of energy today is phenomenally low. Imagine that someone offered to pay you to climb onto a treadmill to generate power to boil water for a pot of tea. How much money would you expect to receive?

Most electric kettles are 3 kW, which is equivalent to 40 slaves. It will take about one and a half minutes to boil with that power input. With only you on the treadmill it's going to take 40 times as long. You will have to tread the mill for one hour to boil the water.

If you get paid the current minimum legal wage, you'll get $5.85 when the kettle boils. Compare that with what you'd pay for electricity generated by fossil fuel or nuclear power—about 10 cents for each thousand watts for each hour. The kettle will use $3 \times 1.5/60 = 0.075$ kWh, costing you about 0.75 cents. So human power, even at its cheapest, is about 700 times as expensive as using fossil fuel.

Even worse, after an hour on the treadmill you will need a change of clothes: you're going to need the washing machine. That means more work on the wheel. There is no escape.

On your treadmill again, it will take twelve hours to heat the water to wash your clothes. And then you're going to need another change of clothes . . .

SPEAK GEEK

— A practical application —

MORE THAN 60 PERCENT OF THE ENERGY FROM BURNING GAS IN YOUR CAR IS WASTED.

There is an unassailable limit to the proportion of the heat energy that can be converted into mechanical power by any kind of engine. The unconverted energy is then dissipated: in the case of a car, out through the radiator and the exhaust pipe.

To figure out how much is wasted, you take the temperature, T_1, of the hot gases made by burning the fuel and subtract the temperature, T_2, of the exhaust gases leaving the machine. Then divide this difference by $273+T_1$ (don't ask—too geeky). Multiply that by 100 to get the percentage of the energy that could be converted to mechanical power. The wasted energy is the difference between 100 percent and the number you just calculated. As a formula, it looks like this:

$$\text{Percentage of energy wasted} = 100 - \left(100 \times \frac{T_1 - T_2}{273 + T_1}\right)$$

For a car engine, T_1 is the temperature of gases in the cylinder just after it has fired, and T_2 is the temperature of the same gases after they have forced the piston down and are pushed into the exhaust pipe. The formula will give you a wastage figure of 60–70 percent; the same limit applies to the turbines in coal, gas, and nuclear power stations. The clouds billowing out of the big towers at power stations are not smoke; they are formed from hot water vapor carrying away the percentage of the fuel's energy wasted as heat.

3

SAFE AS HOUSES

How heavy is your house?

Nowadays we all want to live in our own space. Children leave home in their late teens and want a place to live; newlyweds don't camp with their in-laws; successful people buy houses with offices, double garages, and restaurant-sized kitchens.

Making the materials for building these houses takes a lot of energy. Oft-cited examples of energy profligacy—flying from London to Paris, leaving your TV on standby—pale by comparison.

Building materials such as metal and bricks and concrete that are preformed into a useful shape are processed with heat. Take concrete. The key ingredient of modern concrete is cement, a processed form of limestone and ash. Mix it up with some water, sand, and stones, pour into a hole in the ground, and wait a few days. The slurry changes into a solid of enormous compressive strength that will support a skyscraper or a highway bridge.

It sounds benign. But note that word "processed" preceding "limestone." The limestone has to be heated in a furnace to change into a substance that will combine with water to form solid concrete. The furnace is fired by a fossil fuel, usually gas or coke, and releases CO_2. Worse still, heating causes a chemical change in the limestone that releases even more CO_2 from the stone itself.

Manufacturing a ton of cement pumps about three-quarters of a ton of CO_2 into the atmosphere. That's the amount of CO_2 produced by burning 450 pounds of carbon. If you had to go and buy the carbon as sacks of coal from your local shop, they would fill all the seats in your car, leaving only room for you to sit in the driver's seat.

The story is similar for bricks. Brick is made from clay

fired in a furnace to make it strong and hard. Lots of energy goes into the furnace to get a usable building material, although with bricks no CO_2 releases from the clay itself.

Wood, used all over houses, is light and cheap, and is ideal for structural components such as rafters and joists. It can be shaped by a hi-tech, computer-controlled wood mill in a factory, or by a man with a half-inch chisel and wooden mallet. Most appealingly, it requires little energy to get it to its final shape.

The wood in a tree is made almost entirely of the chemical elements hydrogen, oxygen, and carbon, the last two of which come mostly from CO_2 absorbed from the air around the tree. A solid tree is made literally just from thin air and water.

A living 60-foot pine weighs up to a ton, so each tree ties up a lot of CO_2. You might think that building with wood would deplete this natural CO_2 sink. But take a walk around your local lumberyard yard. You'll see that the lumber has its country of origin stenciled onto the side of the stack and the ends of each piece of wood. Many countries plant more trees each year than are felled, so using their wood encourages reforestation.

What other materials are used in building a house? There is lightweight stuff like copper cables and pipes for the electrical and water systems. Making copper uses energy, but the copper in your house doesn't weigh much, so it doesn't contribute much to the building energy account.

The last significantly heavy and energy-intensive material in houses is steel. Steel rods may reinforce the concrete foundations and in the cast concrete lintels over windows and doors. Steel beams are often used to support floors and roofs.

Steel is no exception to the rule about heavy processed materials using a lot of energy. Changing iron ore into iron and steel requires heat. Most of that heat energy comes from coke and electricity. To produce one ton of steel from ore takes about 3,000 kWh of energy. That's the energy you would get by burning about half a ton of coal, enough to keep your house warm for a month or two in winter.

The weight of a house is a good indicator of the energy used in its construction. So, how much does *your* house weigh? The answer will be about 100 tons, with two-thirds of that in the form of energy-intensive concrete and brick.

But how do you go about weighing your house? You could use a tape measure to get accurate dimensions and then work out the weight of each bit. But it's more comfortable to close your eyes, hold an image of your own house in your mind, and do some rough estimating from your armchair.

A useful mental tape measure is the length of a car. A small car is about 12 feet long, so judging how many cars could be parked along a wall provides an instant way of estimating the width and depth of a building. A similar mental trick to get building heights is to visualize how many men 6 feet tall would need to stand on top of each other to reach a given point on the building.

Start by estimating the outside dimensions. You might be able to park three cars bumper to bumper along the front of your house, and two cars along the side. That would make the front about 36 feet long, and the side 24 feet.

The height of the walls will be about the same as two-and-a-half men standing on each other's shoulders. That would put the top man's waist level with the guttering, making the wall height 15 feet.

Now, with all the measurements in your head, the weighing can begin. Most houses are built on strip foundations of concrete. The strips are laid under every load-bearing wall in the building—the external walls, and also probably a dividing wall running up alongside the stairs.

The concrete is poured into foundation trenches about 2 feet wide, and its thickness will be about 12 inches, or 1 foot. Putting those numbers together, it's safe to say that at least 2 cubic feet of concrete is used for every 1 foot length of foundation strip. So, if the total length of load-bearing wall in the house were 150 feet, the volume of concrete in the strip would be 300 cubic feet.

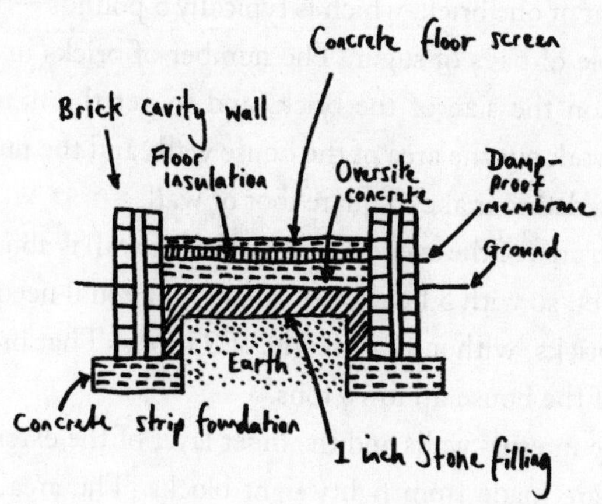

That's not the end of the concrete. The entire ground area of the house within the footings will be covered with concrete to form a base for the floor. On top of that are waterproof membranes, thermal insulation, and a cement screed, making the total concrete thickness in the floor about 6 inches. Its volume is the total thickness multiplied by its area, typically 400 cubic feet. That makes about 700 cubic feet of concrete in total, with a weight of about 45 tons.

Then there are walls. Modern houses have exterior cavity walls. That means an outer and an inner layer separated by a gap of around 2 inches. The gap helps to stop heat from escaping through the wall, and nowadays is usually filled with thermal insulation. The outer layer is often made from brick, and the inner layer from lightweight block. It's normal building practice to use blocks for the internal walls, so we'll count them separately.

The weight of the bricks is the number of bricks times the weight of one brick, which is typically 5 pounds—the same as a couple of bags of sugar. The number of bricks in a house depends on the size of the brick, and to get the number we need to work out the area of the house walls and the number of bricks needed to make a square foot of wall.

The area of the average external house wall is about 2,000 square feet, so with 5 bricks per square foot you'll need around 10,000 bricks, with a total weight of 25 tons. That brings the weight of the house up to 70 tons.

The internal walls and the inner layer of the external cavity wall are made from lightweight blocks. The area of these walls depends on the layout of the house, but you won't go far wrong if you assume that the wall area built out of blocks is about the same as the external wall, but about half the weight. That would add another 10 tons, bringing its total weight to around 80 tons.

Now, in your mind's eye, climb up on the roof and take a look. If you have a tile roof, it might be made of slate, clay, or concrete tiles. Many modern houses have concrete pantiles, but you'll have to check your own roof. A rule of thumb is that there are about 25 pantiles per square foot of roof, each weighing 5 pounds; that's 10 pounds per square foot. If the roof were flat, its area would be the same as the ground area of the house, but real tiled roofs have a pitch angle of at least 30 degrees to stop rain from blowing up under the tiles. That puts the roof area up by about 25 percent, to around 1,000 square feet, and the weight of all the tiles will be about 5 tons.

Now you need the weight of lumber and steel. Most big pieces of wood in a house are used for the floor and joists and roof rafters. They are usually spaced about 16 inches apart, so if you know the length of the house, the number is easily calculated. A figure of 30 is typical. The sizes of the joists and rafters range from 6×2 inch to 10×2 inch. Taking an average figure gives a volume of wood of about 200 cubic feet, which is about 5 tons in weight.

Altogether, the total weight of the house stands at 90 tons. But every builder knows that more material is needed than we have so far counted. It has to do with the fiddly corners, stubs, and overhangs that aren't shown on the drawings. Increasing the weight estimate by at least 25 percent will not be too much. That gives us a total for the house of around 100 tons.

So, if you're now alarmed about all those carbon-killing building materials, you may be wondering how many vacations could you offset by living in a tent. We'll make the numbers easy by assuming that your house has the equivalent of 100 tons of concrete in its structure. Making the concrete will release around 75 tons of CO_2 into the atmosphere. That's the same as 150,000 pounds.

When you fly off for a weekend break, the plane will burn aviation fuel at about the same rate as when you drive your car: 25–30 miles per gallon. Of course, that's just for you and your baggage; the plane actually does about 0.2 mpg, but fortunately it holds more people than your car, so the consumption per person is much less.

Flying a round trip of say 2,000 miles will use 80 gallons of fuel on your personal account, and each gallon of aviation fuel burned releases about 20 pounds of CO_2. Overall, your return trip will put approximately 1,600 pounds into the atmosphere.

Would you forgo a new house for a tent so that you can have 150,000/1600=94 weekend breaks and still hold your carbon-neutral head up high?

SPEAK GEEK

— *A practical application* —

An aesthetically pleasing window has a height 1.618 times greater than its width.

Some houses look right and some look wrong. It often depends on the ratio of the length and width of the building's facade, and of the windows and doors and their positioning in the walls.

To many people, windows and other rectangular shapes look "right" only when their width-to-height ratio is close to 1.618. This number is based on a mathematical rule called the "golden section," which has been used by everyone from the ancient Greeks to Frank Gehry. Using the golden section ensures an eye-pleasing harmony in the shape of the window.

Check your house-harmony by measuring the height and width of its window frames; divide the numbers, and give the result to your real estate agent.

SPEAK GEEK

AN ARCHITECT [illegible] THE WAY [illegible] MORE THAN ITS WORD

4

WELL CONNECTED

Do you know the Queen?

When I was about seven or eight, I went to play with my elder brother on a train track. My brother, deliberately trying to scare me, announced that he was going to derail the next train by placing a penny on the track. I believed him and silently prayed for the imminent disaster to be averted somehow. Penny in place, we hid at the bottom of the embankment. The sound of the steam locomotive grew louder as it approached. There was a terrifying sound of steel on steel, of the snorting smoke box, and the train arrived . . . and passed by. There were no newspaper headlines. The only evidence of our misdemeanour was the penny, now a flattened piece of copper about two inches by one inch, with a slight curl, lying quietly on the ballast between the rails. To this day I still feel a tingle of relief.

I tell this story because we, like other children of our generation, were pretty free of adult supervision. There were pedophiles, transformer yards, and train tracks waiting to catch the unwary child, but the sense of anxiety about strangers and the world's danger barely registered.

Today's unease may be due to not knowing who is who in our community. In a village of three hundred people a hundred years ago, you would probably know which men had a predatory eye for children. What's more, you would know their names, and their parents and their brothers and sisters. Nowadays, you might not even know who lives two doors away from you.

If we look at the number of children born to each couple and do some simple arithmetic, we can see why this has happened, and consequently why we are now faced with the prospect of identity cards, biometric devices, and databases with information about our DNA.

The story goes like this.

How often has someone you know done some name-dropping by telling you about a friend of a friend who has met someone famous, or an eighth cousin twice removed who's starred alongside Tom Cruise? Maybe you know someone whose friend has a relative who works at Buckingham Palace and has spoken to the Queen.

I come from a middle-class family, but went to a secondary modern school. I have no obvious links to the aristocracy. But now I come to think of it, I do know a woman who was married to a man whose father was General Montgomery's adjutant during the Second World War, and who as a result was invited to tea with the Queen. That's just four links between me and the throne.

Try to forge those links for yourself, and you'll almost certainly succeed in finding a connection. In fact there's a good chance of making a chain of personal connections between you and anyone else on Earth using no more than six links in the chain. That means that a friend of a friend of a friend of a friend of a friend knows anyone you care to name—George Bush, Naomi Campbell, Osama bin Laden . . .

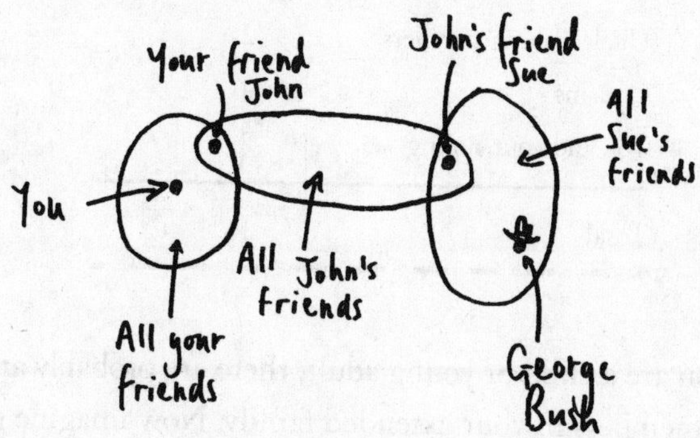

Osama bin Laden probably doesn't have friends that he meets in the local bar, but he certainly has lots of other personal relationships that may eventually lead back to you.

That anyone on the planet is at most only six links away in a chain of relationships seems unlikely until you think about how many people you know. It's simple to estimate the total by dividing them into a few broad groups, which can be enumerated. For example, family, friends, casual acquaintances, and work colleagues probably cover most of the people you know by first name.

Start with an average family in which parents have about two children and assume that grandparents, uncles, aunts, and cousins are included in the extended family.

RELATIVE	NUMBER OF PEOPLE
Your mother's parents	2
Your father's parents	2
Mother and her sibling	2
Father and his sibling	2
Uncles'/aunts' partners	2
Cousins	4
You and your sibling	2
Total	16

So, if you are a child or young adult, there are probably around sixteen members in your extended family. Now imagine going back two or three generations. A century ago, most couples had four or five children who survived into adulthood. It makes the table look dramatically different:

Relative	Number of people
Your mother's parents	2
Your father's parents	2
Mother and her siblings	5
Father and his siblings	5
Uncles'/aunts' partners	8
Cousins	48
Your siblings	4
Total	74

In societies with large families, cousins multiply almost exponentially with family size, and it is they who most likely provide the connectivity that holds communities together.

The reduction in the number of children in each family has had a profound and mathematically inevitable effect on community. In a small town of 10,000 inhabitants at the end of the nineteenth century, only a few hundred extended families would account for the entire population. You, or someone in your extended family, would be related to every other person in town.

Nowadays, with just sixteen members in an extended family, you would have to live in a village of fewer than 200 geographically and socially immobile inhabitants to get the same level of coverage. The tangible result of smaller families is low social connectivity and loss of community.

Anyway, let's get back to analyzing your connection to Osama bin Laden or the Queen. We've established that a typical extended family has sixteen members, or fifteen other people you know well. How about work associates?

Assuming that they are people you know by first name, the number will depend strongly on the kind of organization you work in. Self-employed people might know just half dozen customers by first name, whereas someone working in a large insurance office might know forty or fifty. We could take a typical number as twenty-five.

And then there are your friends. You put in a figure for yourself. I think I have about twenty friends who have nothing to do with my work.

So far, the total number of people with whom you have a sufficiently close relationship stands at about sixty. That figure doesn't include all the casual acquaintances whose name you might know: the man at the post office, the window cleaner, the man you meet walking his dog. Let's say that makes another twenty.

The total number of people with whom you are on first-name terms now stands at eighty. Each of those people in your social circle is at the center of his or her own personal social circle of about eighty people, and each of those people is at the center of his or her circle. To simplify, I'll use "pal" to denote anyone who is a friend, a relative, a work colleague, or an acquaintance.

As a start, we can say that the number of pals of pals we have is about $80 \times 80 = 6,400$. But in real life some of the pals'

pals will be members of your family, work colleagues, and circle of friends, and the figure of 6,400 counts those people twice.

A simple way to take that overlap into account is to introduce a factor that expresses the percentage of each successive social circle common to the previous group. A factor of 50 percent seems fair. It means that of the eighty people in the social circle of one of your pals, forty will also be in your own circle.

So now, very conservatively, each link in the chain of social circles multiplies the number of people connected to you by a factor of just 40, but successive multiplication builds up the numbers quickly. Try it out for six links in the chain:

Links	Number of people
1	40
2	40×40=1,600
3	40×40×40=64,000
4	40×40×40×40=2,560,000
5	40×40×40×40×40=102,400,000
6	40×40×40×40×40×40=4,096,000,000

There are 4.1 billion people accessible through a chain of just six links. The world's population is about 6.6 billion, so it seems that a pal of a pal of a pal of a pal of a pal will connect you to three-quarters of humanity.

Does it really work for anyone? Finding a link to Osama might be tricky, although certainly of interest to the CIA. Bin Laden is a Saudi citizen, so a good bet would be to look for a link between yourself and someone in Saudi Arabia. Once you have found that link, there are likely lots of links between that person and someone in bin Laden's family.

These links will connect you to almost anyone on the planet, but remember: it takes only one suspicious individual on a chain to cast suspicion on all the other people in the chain—the Queen, perhaps?

SPEAK GEEK

— A practical application —

A TEN-DIGIT PHONE NUMBER COULD CONNECT YOU TO ANYONE ON EARTH.

Have pity on small-town undertakers. In the 1880s, an undertaker called Almon Strowger had a problem. His rival's wife was the local telephone operator, and she put funeral calls through to her husband instead of him. Strowger realized that her family loyalty spelled death for his own business and set about inventing an automatic telephone switching system.

His simple idea remained in use until the 1980s. The wires from a telephone were connected to a switch with ten positions. Each of the ten outgoing connections from the first switch was connected to another ten-way switch. The switches were operated automatically by electrical pulses generated when the telephone dial was turned.

After just one 10-way switch it would be possible to select one of ten other phones to talk to. After two switches it would be $10 \times 10 = 100$, after three switches it would be $10 \times 10 \times 10 = 1{,}000$.

Theoretically it would be possible to have a single Planet Earth Telephone Exchange. With ten 10-way switches connected together, you could be connected to any one of $10 \times 10 \times 10 \times 10 \times 10 \times 10 \times 10 \times 10 \times 10 \times 10 = 10$ billion other phone users.

But with ten of those switches for every phone user on

the planet, the telephone exchange building would be enormous. So, in practice the switching is split up into millions of little switches covering small areas. It means that more than ten digits are needed, but thirteen will still connect you to nearly anyone.

Now you really are well connected . . . by switches connected to switches connected to switches.

5

FATAL ATTRACTION

How much are you physically attracted to your partner?

Have you ever woken up in bed with a numb arm or leg? The blood flow to the affected limb is constricted and leaves it feeling deadened. So you change position, the blood circulates again, and you go back to sleep.

The weight of a dead leg or arm seems enormous as you drag it into a more comfortable position. But how much do your individual body parts actually weigh? Your whole body weight is easy to measure: you just step onto some scales. Weighing a leg, an arm, or even your head is more challenging, assuming that your body is to be left intact in the process.

Here is a practical method of working out an approximate value for the weight of your head while it is still connected to your body. You'll need a bucket large enough to immerse your whole head in, and a kitchen measuring bowl. A bathroom towel might make the experiment more pleasant. It is best to do it outside, but be careful not to alarm the neighbors.

Fill the bucket to the brim with water and place it on the ground. Kneel next to the bucket, take a deep breath, and gently immerse your head in the water, up to your neck. Water will spill out of the bucket. This is quite normal.

When you've surfaced, the bucket will no longer be full. In fact, the amount of water that has overflowed will be exactly equal to the volume of your head. You can get a good enough measure of that volume by topping up the bucket with water using the measuring bowl, until it is again full to the brim.

I just did this; you might notice the wet pages. The volume of my head worked out to be precisely 1.3 gallons. Now, I wouldn't like you to think that I'm being big-headed about this, but that makes my head about the same size as one of those plastic gas containers you keep in your car trunk in case of emergency.

Hopefully you've also done the experiment and will now know the volume of your own head in the units of your measuring bowl—liters, pints, or fluid ounces. If the stuff in your head had the same density as water, the weight of your head would be equal to the weight of the same volume of water. A pint of water weighs 1 pound, so the weight of your head in pounds would be the same as the number of pints of water. In my case (and for most adults) that'd be about 11 pounds.

Despite appearances, your body is mostly water. But compared with other parts of the body, a head contains quite a lot of bone relative to soft tissue, and so is going to weigh more than water, maybe 10–20 percent more. So just increase your initial head weight estimate by this amount. That brings it up to thirteen pounds.

In scientific terms, weight—whether of your head, your whole body or anything else—is a force that pulls between the mass of an object and the mass of the earth. If you step onto old-fashioned bathroom scales that work by compressing a spring, the force that is your weight compresses the spring as surely as if you had squeezed it with your hands. What gives rise to that force is, of course, gravity.

Sir Isaac Newton worked out that the strength of the gravitational force pulling two objects together increases in proportion to both masses, but decreases rapidly as the distance between the masses increases. He showed that the force was quartered for each doubling in distance.

We can write this as a formula:

$$\text{Force} = G \times \frac{\text{Mass}_1 \times \text{Mass}_2}{\text{Distance}^2}$$

That symbol G is called the *universal gravitational constant,* and if the force, masses, and distance are measured in pounds and inches, it has a value of about $4.7e^{-9}$. The number written out in full is 0.0000000047, but it is very tedious to write down all those zeros, so it's easier to use the shorthand. The –9 after the e indicates that the digits 47 are preceded by a decimal point and 8 zeroes.

The first person to devise a method to get an accurate value for the universal gravitational constant was Henry Cavendish. In 1798 Cavendish, reputedly a shy man who may have had Asperger's syndrome, set up an experiment using an instrument called a torsion balance.

The torsion balance is a wooden rod suspended at its midpoint by a silk thread. Weights are hung from each end of the rod, like a mobile above a baby's crib. Cavendish sealed the delicate mechanism in a room so that it would not be disturbed by draughts.

He moved large balls of lead close to the outside walls of

the sealed room, causing the suspended rod to twist slightly under the influence of this gravitational force. The amount of twist can be used to work out the force and hence the universal gravitational constant.

That same gravitational force pulls on your body whenever there is a nearby massive object, be it a ship, a mountain, or another human being. Unlike the pull from the earth, those forces act along a line between you and the other object.

For example, when you stand next to your partner, how much are you pulled together by the gravitational force between your bodies? Let's make it intimate—say you and your partner are standing face to face, just lightly touching.

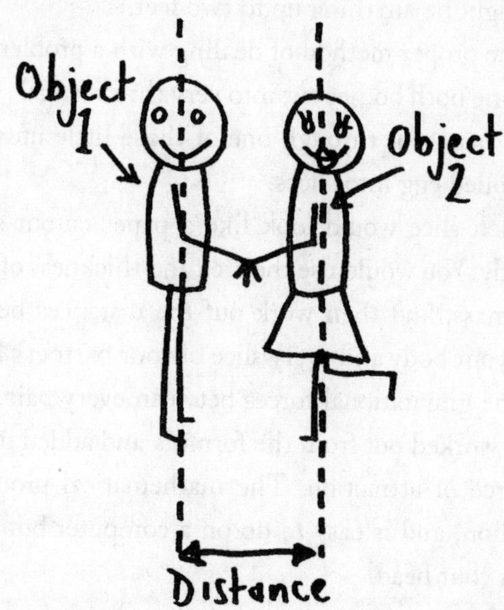

To work out the strength of your attraction to your partner, you'll need some numbers to plug into the gravitational force formula.

Say that $Object_1$ is you. What is your mass? In numbers, it's the same as your weight in pounds. This is just an estimate, so use a nice round number, say 200 pounds. Next you need the mass of $Object_2$. That's your partner, and it's up to you to negotiate a value to insert here, but the calculation will be easier if you use 200 pounds again. It may, of course, make other parts of your life more difficult.

Last, to use the formula you need a value for the distance between your bodies—and this is the interesting bit, because no single distance is correct. The distance at the points where your bodies touch, say, umm . . . chest to chest, is almost zero, but the distance between your buttocks and your partner's buttocks might be anything up to two feet.

The proper method of dealing with a problem like this is to imagine both bodies cut into very thin vertical sections, as if you had been put through one of those little utensils for cutting a boiled egg into slices.

Each slice would look like a paper cutout silhouette of your body. You would use the area and thickness of each slice to get its mass, and then work out the distances between every slice of your body and every slice of your partner's body.

The gravitational forces between every pair of slices can then be worked out from the formula and added up to give the total force of attraction. The mathematical process is called integration, and is easy to do on a computer but too complex to do in your head.

What we need is a rough and ready method that will tell you quickly how much you are attracted to each other. He or she may be waiting impatiently to know the answer.

A simple solution is to assume that for you and your partner, all of your mass is concentrated at a single point in the middle of your body, say halfway between your belly button and the small of your back. That might be around 6 inches in from your front, making the effective distance between your body masses 12 inches.

And now you can run the formula. These are the values we need to plug in: $Mass_1 = 200$ pounds, $Mass_2 = 200$ pounds, and Distance = 12 inches. So your mutual attraction will be:

$$Force = 4.7e - 9 \times \frac{200 \times 200}{12 \times 12} pounds\ force$$

That's a force of about 1 millionth of a pound force.

Newton, strangely, received his knighthood not for his scientific advances but for his work as Warden of the Royal Mint. He stabilized the currency by linking the value of silver coins to the value of gold—a kind of precursor to the gold standard, used off and on until the Second World War.

He probably wasn't much interested in the kind of inter-body attraction we are calculating. He seems to have had only one girlfriend, whose affections dimmed as he immersed himself in science.

That's a tiny force between you and your partner. Maybe Newton had already calculated that his force was too small to hold people together.

It is a measurable force, but you would be well advised to rely on the force of your personality, or other attractions, to keep your partner close.

SPEAK GEEK

— A practical application —

A man who weighs 200 pounds at the North Pole would weigh 2.5 ounces less at the equator.

Your net weight is the result of two forces. The first force is gravitational: it pulls your body mass downward, toward the center of the earth. The second force is upward, and is caused by the rotation of earth constantly trying to fling your body into space like a stone on a string being swung around your head.

At the exact point of the North Pole, earth experiences no rotation, and so there isn't any upward force. On the other hand, earth's rotational speed at the equator is at its maximum—about 1,000 mph—and a small amount of the gravitational pull is canceled out by the upward flinging force. Hence the weight loss.

Beat *that* for a diet, Professor Atkins.

6

HOME ALONE

How many piano tuners are there in Boston?

How much do you have in common with a Bushman in the Kalahari Desert? You both need food and shelter, shade from the sun when it's hot, and a source of heat for warmth when it isn't. Like him, you nurture your children. On the other hand, how often does a Bushman peruse the menu in a cappuccino bar?

We aren't all the same, of course, but each of the different societies that exist across the world can be placed in one of a fairly small number of distinct categories. Western industrial, agricultural, developing—these are labels for countries that provide us with a convenient picture of a particular social, economic, and cultural behavioral pattern.

Within those broad categories there are finer divisions that are culturally specific and instantly understood by people belonging to a particular cultural group. Try dividing the list of words below into two groups with a common thread in each group:

Tuscany	Detroit
satellite TV dish	pinstriped
Club Med	Greenwich
opera	tennis
public school	minimum wage
pearls	layaway
Financial Times	director
football	salary
accordion	low-income housing

Here is one obvious division based on social group:

public school	salary
football	pearls
satellite TV dish	opera
accordion	Tuscany
Detroit	*Financial Times*
minimum wage	tennis
low-income housing	director
layaway	pinstriped
Club Med	Greenwich

We have strong preconceptions about the attributes of different social or lifestyle groups. They may be rooted in actuality, or be unfounded prejudice, but we use them all the time to simplify the complexity of life.

In consumer societies, social labels are often in terms of ownership rather than beliefs, income, or occupation. Here's a sample of objects whose ownership is routinely used to put people into a nice tidy pigeonhole:

SUV	rowhouse
yacht	piano
hybrid car	garden
bicycle	hunting gun
second home	dogs
RV	horses
house	swimming pool

Think about the things you own. How would you be labeled? For example, I play a piano accordion, and a lot of my friends don't regard the instrument as musically respectable. But probe a bit deeper and it becomes obvious that their judgment is more about association: the accordion is not a "middle-class" instrument. And, looking around at other accordion players, it's true that many come from working-class backgrounds.

Accordions are portable, don't take up space like a piano, and are relatively inexpensive—all things that matter if your income is low, your housing uncertain, and you have to follow your job. And the accordion is the immigrants' instrument, allowing them to take a bit of the old culture to their new country.

With pianos it's just the opposite. The piano is, after the violin and the cello, the most socially respectable of all instruments. Having a piano in your front room is a mark of intellect, aspiration, and culture.

Social grouping and owning certain things go hand in hand. If you know the social group, you'll be able to predict the possessions; if you know the possessions, you can predict the social group. This relationship can be used to predict quite a lot about people and objects in places that you might have never even visited.

Take Boston, Massachusetts, featured in TV shows such as *Cheers*.

All in all, the social groups, value system, and wealth of Boston are going to be similar to those of any moderately affluent city in a developed nation, and the similarities allow you to make statistical predictions about the objects owned by people in Boston.

For example, if you knew the number of pianos or accordions owned by people in a city such as London or Singapore, you could directly estimate the number in Boston by comparing the population of Boston with the population of Cambridge or Liverpool.

That's quite easy. Now try using the comparison method to predict a slightly less obvious quantity. Can you work out the number of *piano tuners* in Boston? Can you do it without looking in a book, phoning a friend, using Google . . . ? In fact, can you work it out sitting by yourself at home in your chair with your eyes closed?

Even if you have been to Boston, the chances are that you weren't there to hire a piano tuner. But you can still make an estimate. All you need to do is analyze the statistics of piano tuning where you live, and then apply what you've found to a city the size of Boston. Of course, the estimate will be poor if the locality in which you live is very different from Boston; for example, the ownerships of pianos in downtown Milwaukee will not be typical of the United States as a whole, let alone Boston. You need to allow for such variations.

Start with your own piano encounters. When you went to school you were probably in a class of about thirty children. Ask yourself: how many of those children had a piano at home?

It's likely that there was at least one kid with a piano. On the other hand, it's not very likely that more than half a dozen children spent their evenings tinkling the ivories. Taking a middle figure, you might estimate that three out of thirty families have pianos. But this applies only to families with children at school.

When children grow up, some parents will get rid of the piano. So, if you want to extend your estimate of piano ownership to all households, you should probably reduce the original figure. Let's say 1.5 pianos per thirty households instead of three pianos per thirty families—or in whole numbers, one piano per twenty households.

Now it looks easy. Just assume that the pattern of piano ownership is similar in Boston, estimate the number of households in Boston, and divide that by twenty. Voilà—the number of pianos in Boston. But how do you estimate the number of households in Boston when you are sitting in a chair with your eyes closed?

We could get at this from a figure for the population by assuming that each household has an average of, say, three people. The number of households is then the population divided by three. Divide this number again by twenty to get the number of pianos.

All simple so far, but we still need a population figure. You could just look it up, but the rules of this exercise forbid getting out of your chair. So the trick is to think of a city whose population you know, and that has some point of similarity with Boston.

For example, I as a Brit know that Cambridge, England, has one major university and a population of about 110,000. I also know that there are at least three major universities—Harvard, MIT, and Northeastern—in the Boston area.

So maybe we could estimate the population of Boston as four times that of Cambridge, on the grounds that it has four times as many large universities. That would make our initial population estimate 440,000. This number might be right, but it's a shaky estimate, based as it is on just one point of similarity.

The evidence for our initial estimate is so weak that we really need to look for more reference points to back it up. It ought to be something about Boston you know to be similar to a city you know.

So here's another angle: Boston is an Atlantic seaport, like Liverpool. I know the population of Liverpool—it's 440,000. In the past, much of the shipping between the United States and Britain was between Boston and Liverpool, so it's quite likely that their sizes are similar. In fact, the population estimate of 440,000 ties in roughly with our first estimate. (Actually, I cheated and looked it up; the population of Boston is nearer 600,000, so our estimate is not too bad.)

Back to counting pianos in Boston. First, we need the number of households. It's going to be 600,000 divided by 3 people per household. That makes 200,000 households. According to our earlier mental calculation, 1 in 20 households owns a piano. So that makes 10,000 pianos.

The final step is to work out how many pianos can be serviced by a single piano tuner before they need to be tuned again. Dividing that figure into the number of pianos will give us the number of tuners.

If you have ever lived with a piano, you'll know that a tuner might cross your threshold once a year at most. That is, once every 365 days. To make the arithmetic simple, let's say that there are 200 working days in a year. A piano tuner spends a couple of hours with the piano, and takes maybe an hour between jobs, so each piano takes three hours. But then there's lunch and things, so a tuner can do just two pianos a day. So, in the 200 working days between two tunings of your piano he will have tuned a total of 200×2=400 pianos.

We estimated that there are 10,000 pianos in Boston, so it's going to take 10,000 divided by 400 tuners to get round them all before they need retuning. That's twenty-five tuners.

Are you still sitting in that chair with your eyes *closed*?

SPEAK GEEK

— A practical application —

More than 30,000 elephants were slaughtered just to make ivory piano keys.

The ivory veneer on Victorian piano keys is about 1/16 of an inch thick and approximately 8 inches long by 3/4 inch wide. That amounts to 3/8 of a cubic inche of ivory per key.

There are 58 white keys on a piano keyboard, bringing the total volume of ivory per piano up to 21 cubic inches. It's a reasonable guess that there were up to 10 million pianos worldwide in the Victorian era. The total volume of ivory needed in their construction would be 210 million cubic inches, or 180,000 cubic feet.

A typical pair of adult elephant tusks has a volume of about 3.5 cubic feet, so at least 34,000 elephants were killed for the ivory in our forefathers' pianos. What's worse, it's likely that many more were killed as not all of a tusk yields usable ivory.

7

BEAM ME UP, SCOTTY

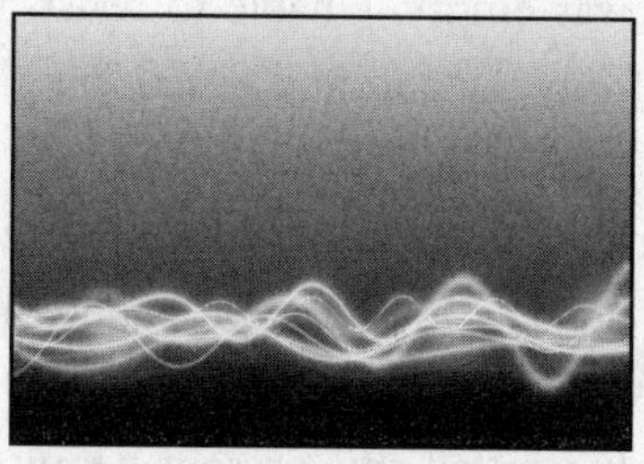

How long would it take to send your body to Mars?

Captain Kirk has been sorting out a spot of bother on some planet or other, while the *Enterprise* orbits at a safe distance. A radio command is sent to the ship: "Beam me up, Scotty." His body fades out, then reappears, shimmering, on the starship.

It seems to happen so quickly, but how long would it really take? This is not just about the finite speed of light. It's much more subtle, and to understand it you need to know about my father's coffee flask.

It was standing on the kitchen table. It was 1962, my birthday, and my father had just returned from work. I knew that the flask contained a treat that had been promised for weeks.

The room was empty. Unable to resist, I unscrewed the aluminum cup from the top and eased out the cork stopper. Instantly, white vapor clouds poured over the top of the flask and rolled along the table top and down to the floor.

This was liquid nitrogen—probably the best birthday present I have ever received.

I desperately wanted this present because of the satellite receiving station at Goonhilly Downs, in Cornwall. The station had recently been commissioned, and everyone had watched the first flickering live TV pictures to arrive from America via the Telstar satellite.

Getting the signal from the ground to the satellite was not a problem. The TV pictures were beamed up using many kilowatts of radio power, focused onto the satellite by a giant parabolic dish antenna, like sunlight through a magnifying glass.

The difficulty was getting the signals back down to earth. The satellite relied on small solar cells for its power, so it could generate only a few tens of watts of radio power—about as powerful as a car headlight. Worse still, the satellite was so small that it couldn't carry a big focusing antenna, so the radio waves it beamed back were spread out over a large part of the earth's surface, an area called the satellite's "footprint." The footprint was about 600 miles by 600 miles. That's almost half a million square miles. And that's why I wanted liquid nitrogen.

Let's say that the satellite transmits a radio power of 10 watts. That's about the same as the light of ten candles at a distance of 25,000 miles. The task of the ground station is to "see" those ten candles. The power spreads out over the footprint, giving less than ten-millionths of a millionth of a watt on each square yard of ground. That is a very small power.

Now here is the problem. The tiny power is picked up by the antenna of a radio receiver and causes an electrical current to flow in its wires and other electrical parts. An electrical current consists of electrons in motion. But even when there is no signal current flowing, the electrons are still in motion because the atoms to which the electrons are attached are themselves jiggling around. This jiggling generates a small electrical power in wires and other components, and that power, though tiny, is comparable to the radio power collected by the antenna.

And here is the nub of the problem. No amount of amplification will make the radio signal detectable if it is already mixed up with the electrical noise generated in the wires and other components in the first part of the radio receiver. Amplification will increase the noise as much as it does the signal, so the signal is forever swamped by the noise.

But the engineers at Goonhilly had a couple of tricks to deal with this problem. The first was obvious: they got themselves a massive parabolic dish antenna, which they named

Arthur. Arthur had an area of thousands of square yards and greatly increased the amount of radio power they could collect.

But there still wasn't enough signal to make it much bigger than the noise generated by random electron movement in the radio receiver. And thus the second trick came: the engineers cooled all the wires and the electronics so that the atoms in them jiggled more slowly and generated less noise.

The motion of the atoms is caused by the temperature of the material. In fact, what we call the *temperature* of something is just a measure of the average jiggling speed of the molecules or atoms of which it's made. Cool everything down, and the speed is reduced, along with the electrical noise.

The cooling at Goonhilly was done by circulating liquid helium around the main electrical components. Helium boils at about –452° Farenheit, just a few degrees above absolute zero—the temperature at which atoms cease to move—so it's very effective at reducing electrical noise.

Liquid helium is difficult to make. It's certainly too expensive to give to a twelve-year-old as a birthday present, and that's why my father gave me liquid nitrogen instead. Its boiling point is a mere –320° Farenheit, but that's still cold enough to reduce radio noise.

I had planned an experiment to see if I could reduce the background noise in a homemade radio. But first I thought I'd see what liquid nitrogen would do to rubber. I lowered a piece of soft orange rubber tubing into the flask. There was an eruption of boiling nitrogen, and I lifted out the tubing, now as stiff as a rod. I struck it with a hammer, and it shattered into tiny orange shards like a broken flowerpot.

The disadvantage of having older siblings now kicked in. My sister poured some of the liquid over the fruit bowl on the sideboard, causing an explosion of white vapor clouds. Then my brother grabbed the flask and poured liquid over the sleeve of his school blazer. Bubbles momentarily frothed on his arm. The flask was empty. There was no liquid nitrogen left for my experiment.

These days we can easily pick up signals from satellites; we use tiny, low-power cell phones; and we can listen to miniature space probes as they tour the surface of Mars. We do it all without immersing our receivers in liquid helium. So what has changed?

One important factor is brain power. One particular brain belonged to a mathematician-engineer called Claude Shannon, who laid the foundations of a subject called information theory. He published his key academic paper in 1948. Shannon developed his ideas at Bell Labs and MIT.

One of Shannon's insights was that in radio, it's possible to trade *time* for *power*. The power needed to send and receive a signal in a noisy environment can be traded for the time taken for its transmission. He proved this theory with some elegant mathematics, but the idea is simple.

Imagine that you are standing on the sidewalk on one side of the street. A friend is standing on the opposite side. Traffic is moving in both directions, and it's almost impossible to cross. So you shout to your friend, "I'll meet you later." This is your message, but it gets so mixed up with the traffic noise that your friend has difficulty hearing you.

Maybe the first time you shout, she will hear the word "you," perhaps because there was a momentary lull in traffic noise. You shout again, and this time she gets the words "later" and "meet." You shout a third time, but she doesn't catch anything new. On the fourth repeat she hears "I'll."

She can now piece together the complete message without any errors: "I'll—meet—you—later." But the message has had to be repeated four times, so the time taken to convey the information was four times as long as if there had been no traffic noise.

If the traffic had been even noisier, you would have had to repeat the message even more times; less traffic, and you'd need to repeat it only once or twice.

It turns out that message repetition is just a special case of a process called *coding* that can be applied to a message before it's transmitted. There are better coding techniques than repetition. They work by replacing the original message symbols with new symbols that are worked out mathematically. The coded message is longer than the original message and so takes longer to send. But after such coding, it can withstand the background noise generated inside the radio receiver. Cell phones use this kind of coding.

Shannon and another electrical engineer, Ralph Hartley, formulated a scientific law called the Shannon-Hartley law, which puts a number on the bottom line in message coding. The law says that given an ideal coding scheme, the absolute maximum rate at which information can be sent in a message is limited by the strength of the signal compared with the background noise.

Everyone knows that radio waves travel at the speed of light in free space. The Shannon-Hartley law shows that although the waves themselves travel very fast, the speed of the information they carry is limited by the background noise.

So, this is where we can start doing some rough calculations for our matter transference beam. Specifically, how long would it take to send your body via a beam like the ones in *Star Trek*? Of course, such a system doesn't yet exist, but it's useful to be prepared.

There are several possibilities for beaming your body across space. The one I favor avoids invoking extra dimensions, parallel universes, or as-yet-undiscovered quantum mechanical effects. My system uses radio.

A century from now, when you enter the deconstruction chamber at Heathrow Space Port, the structure of your body will be analyzed down to the last cell. That includes the precise structure of each strand of DNA in each cell, as well as the composition of your body at the instant of deconstruction. The deconstruction will probably need to be destructive, so your earth body will cease to exist physically at this point.

But the *information* about your body will still exist, stored temporarily on a computer in the departure lounge. When a channel becomes free, your *body-info* will be sent by radio and stored on another computer in visitor reception area on Mars.

The most important device in the reception area will be the Bio-Constructor, a machine that resynthesizes your physical body based on the transmitted information. It will use proteins and other biological building blocks kept in dispensers on Mars.

How long would it take to transmit your body to Mars—or really, the time to transmit the information about your body structure, your body-info? You would probably prefer this to be done without any errors.

The first thing to determine is how much information is needed to describe your body down to the DNA level. One approach is to say that it's the amount of information contained in the human genome—estimated to be about 6 gigabits. A bit is the amount of information given by saying that something is *on* or *off.* There are 8 bits in a byte, so 6 gigabits is the same as 750 Mbytes—an amount that could easily be stored on a computer. (The prefix "giga-," often abbreviated to "G," stands for multiplication by a billion.)

Some people would point out that the information about our bodies is not stored solely in DNA. A lot of the information is implicit in the properties of the substances from which our bodies are built. That would reduce the information count still more.

But we should probably take the more pessimistic view and reckon on having to send the information about every individual cell in your body in order to re-create it exactly as it was at the instant of deconstruction at Heathrow. There are between 100 billion and 1,000 billion cells in your body. Many of those cells have identical structures, but we would still need to send information about the presence or absence of each cell, along with its type. At the very least it is going to require 1 bit of information per cell, giving a total information count of up to 1,000 billion bits. That's 1,000 Gbits (125 Gbytes).

Now we are ready to use the Shannon-Hartley law to work out how long it will take to send your body-info to Mars. Here's the law written as a formula:

$$C = B \times \log_2\left(1 + \frac{S}{N}\right)$$

The formula says that the absolute maximum amount of message information that can be sent in one second, C, is equal to the bandwidth, *B*, of the radio spectrum that can be used, multiplied by the logarithm of 1 plus the ratio of the signal power, *S*, to the unwanted noise power, *N*.

To use the formula, we need a value for *B*, the bandwidth of radio spectrum available for transmission. One hundred million hertz is the kind of bandwidth used for space probes, so we'll use that figure.

Then there's the ratio of the power of the radio signal received on Mars to the noise generated in the radio receiver. That's the *S/N* part of the formula. This is the same problem that was tackled by the engineers at Goonhilly in 1962. If we use very big parabolic dishes, both on earth to transmit and on Mars to receive, and liquid helium to cool the radio, we might get the signal-to-noise ratio up to about 1.

Putting those numbers into the formula gives the maximum possible number of bits of information that can be sent each second. It works out at 100 million. So, to send your 1,000 Gbits of body-info will take 1,000 billion divided by 100 million, or 10,000 seconds.

That's a three-hour trip, which is about how long it takes to fly from Heathrow to Athens, Greece. The only things to worry about are a computer crash while you are in transmission and making sure that your Dad brings the helium for cooling the radio on Mars.

Oh, and one other thing: the precise pattern of neural activity in your brain at the instant of your deconstruction has not been transmitted. Your body will arrive on Mars, but you won't remember why you are there, or even who you are.

SPEAK GEEK

— A practical application —

It takes 1/4 second for your words to go via satellite from Britain to North America.

Communication satellites sit in geostationary orbits at a height of 25,000 miles. A geostationary orbit is an orbit whose height is carefully chosen so that the satellite takes exactly one day to orbit the earth. Because the period of the satellite's orbit exactly matches earth's period of rotation, the satellite stays above the same point on the earth's surface all the time.

The journey time for the radio signal carrying your words is calculated by dividing the distance the signal travels by its speed. That's around 50,000 miles (sending the signal up and then back) divided by the speed of light, 186,000 miles per second—0.26 seconds each way.

Undersea cable phone links don't suffer from this time delay. Although the speed of the signals along the cable is less than half the speed of light, the distance the signal has to travel is much shorter—only 3,000 miles from Britain to North America. That's a journey time of only 0.03 seconds, not enough to disrupt the flow of conversation. And that's why most telephone conversations now go by cable.

8

HIDDEN DEATH

How much land is needed to bury the dead each year?

Newspaper headlines give the impression that something is terribly wrong, and Something Must Be Done:

FIVE DIE IN HIGHWAY CARNAGE

SUPERBUG KILLS HOSPITAL PATIENT

HUNDRED DEAD IN PLANE DISASTER

And yet every day, all around us, thousands of people are quietly dying, and there are no headlines to mark their passing.

Let's use my own country as an example. In Britain the population is around 60 million. Leaving aside emigration and immigration, that must mean that the number of people born every year is about equal to the number of people who die every year. But just how many people die every year?

You can estimate the figure quite easily from the average lifespan in this country, which is roughly 75 years and rising. Imagine for a moment that all births in Britain stopped today, that from now on people die off and aren't replaced.

Each year more die, until, after the time of the average lifespan of 75 years, there will be very few of today's 60 million people left. That means that about 60 million of us will die in the next 75 years. So, if ages are evenly spread, an average of 60 million divided by 75 people will die every year. That's about 800,000 per year. In reality, the number of people dying every year is quite a lot smaller, around 500,000. The difference between the simple estimate and the real number is because our lifespan is currently increasing quite rapidly. But estimates are all about getting an answer that is *about* right, so we'll continue working with the figure of 800,000.

How does this figure compare with the number of people

killed in natural disasters and major accidents? For example, how many people do you think lost their lives because of the accident at the Chernobyl nuclear plant? Was it 5,000? 10,000? 50,000?

The official report by the United Nations, the World Health Organization, and the International Atomic Energy Agency puts the number of deaths caused directly by the accident at less than 50, but says that up to 4,000 could eventually die prematurely from exposure to radiation. The report not only caused outrage in the environmental movement but also contradicted popular belief that hundreds of thousands had either already died or would die.

Imagine that the accident had happened in Britain, with its population of 60 million and an estimated death rate of 800,000 people per year. Would it be possible to know whether people were continuing to die as a result of the accident? The first 50 deaths would be obvious enough: they would be the heroic firemen who took turns to shovel radioactive debris into the containment area. They would die in the hospital, in the first weeks or even days after direct exposure to the radiation.

Let's take one of the wilder predictions of the Chernobyl death toll: say that 100,000 die prematurely from radiation-related diseases in the years after the accident (a figure 25 times greater than the UN estimate). Surely we'd be able to notice those additional deaths? The ages of all the people affected at the time of the accident are likely to match the overall age profile of the population, from infants to the elderly. The elderly might have only a few years of life left in which to develop a disease from the radiation; their early demise would be hard to pick out from standard mortality statistics. It will be the very young, who had potentially 75 years of life ahead of them, whose early deaths will be most noticeable.

If those early deaths are evenly distributed over, say, the next 20 years, we might expect the higher mortality rate in any year after the accident to be 100,000 divided by 20—5,000 extra people dying each year against a backdrop of 800,000 "normal" deaths. The excess mortality then amounts to less than 1 percent of the average.

But the number of deaths each year will fluctuate because of other factors: a snap of cold weather, hot weather, food poisoning, an increase in national wealth, even social upheaval. So, perhaps surprisingly, it is unlikely that the death count we imagined from a British Chernobyl is going to be discernible in the basic yearly statistics: too many people are dying all the time for the signal from the accident to stand out from the noise of everyday life and death.

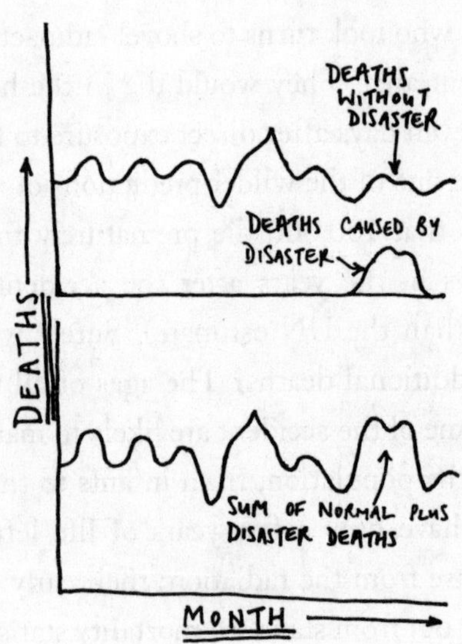

To detect the Chernobyl death signal, we need to get beyond the raw data. For example, if radioactive iodine were released in the accident, the incidence of thyroid cancers would be expected to increase in the years that followed. Look at the number of people with thyroid cancer before and after, and you might well detect the effect of the accident. But if, as the UN says, only 4,000 people will die over the following decades, it may still be impossible to spot any correlation with the accident, even with this "data filtering." There are simply too many people dying of other causes for it to make a noticeable difference.

On a lighter note, the bodies of those who die each year from whatever cause have to be disposed of. Some will be buried, some cremated. The scale of deaths is easier to imagine in terms of graveyard plots, and it's easy to calculate.

You need to know the proportion of people who get buried as opposed to cremated. You also need to know the average area required for a burial. Think about the funerals you have been to. How many were burials and how many cremations? You'll probably find that in the West it's about 30 percent burials and 70 percent cremations: more cremations in cities, more burials in the country. If 30 percent of those who die each year are buried, the number of burials is going to be about 30 percent of 800,000, which is 240,000 per year.

To get the total land area you need to estimate how much area each burial uses. In a British cemetery, graves are in rows spaced about 6 feet apart, and the rows are about 12 feet apart. That means that each grave requires about $12 \times 6 = 72$ square feet.

Multiply the grave area by the number of burials and you

get $72 \times 240{,}000 = 17{,}280{,}000$ square feet, call it 20 million square feet. It might help to imagine that area as a number of soccer fields. The average size of a field is about 325 feet long by 160 feet wide—an area of 50,000 square feet. We need the equivalent of 400 soccer fields to bury our dead each year.

Each year, you'd need about one-third of a soccer field to bury all the people who die in the United Kingdom as a result of the Chernobyl accident.

SPEAK GEEK

— A practical application —

A STATISTICIAN MIGHT NOT NOTICE A DISASTER IN WHICH 200 PEOPLE DIE.

In Britain about 36,000 people die in each of the summer months, and the monthly tally rises to 48,000 in the depths of winter. These are average figures. The actual monthly totals fluctuate around this underlying seasonal trend. The fluctuation tops out somewhere near 200 deaths and is called, rather portentously, the "de-trended standard deviation."

For this reason, a disastrous accident causing 200 additional deaths in one month would not necessarily show up in monthly mortality figures. A statistician scanning the figures wouldn't know if the blip was natural fluctuation, or a disaster.

9

IT'S FOR YOU

What is the most surprising message you will ever receive?

There's a tiny hamlet called Gallanach on the west coast of Scotland. Built into its rocky shore is the station for the transatlantic cable that is Britain's telecommunications link to the New World.

I visited this place in short trousers with my father in the early 1960s. He always arranged a special visit for us children during our summer holiday. I thought it fantastic, but my brothers and sisters groaned when each trip was announced.

At Gallanach, the cables duck under the concrete floor at the start of their underwater journey across the rifts and mountains of the Atlantic Ocean. Inside the station there is a faint background sizzling sound from the electronic clocks that control the digital traffic. Once outside, the cables snake beneath the sand before plunging under the sea toward America.

One of the most famous cables was known as TAT-1. During the Cold War it carried the "Hotline" between Moscow and Washington, and more than once our future depended on its reliable operation. The cable became operational in 1956, although the Hotline agreement between the United States and the USSR was not signed until 1963, after the Cuban missile crisis had taken the world to the brink of nuclear war.

Phone messages between Harold Macmillan and JFK during the crisis would have been routed via the Gallanach cable. But what exactly is a message? We can think of a message as a communication that conveys new information.

Imagine that Kennedy had rung Macmillan and said that he was about to launch his missiles. That's undeniably a message, and it would be news for London. There would be warnings on the radio, sirens sounding, white-faced adults and crying children crowding into shelters.

Just how much information would there be in such a message? Information is measured in units called bits, just as the length of a piece of string can be measured in inches.

Put yourself in Macmillan's position as prime minister. In a time of international crisis, he would be getting regular calls from the White House. One night the call arrives, and JFK says, "This is it. I'm going to launch a preemptive strike against the Soviets."

Now, if Macmillan had already been certain that this was going to happen, Kennedy's message would provide no new information. By our definition, therefore, the information in that message can be said to contain zero bits. Messages that say something you know already are worthless.

But what if Macmillan had been thinking that the odds of the launch were even? Kennedy's message makes it certain. It wipes out the 50 percent uncertainty that existed in Macmillan's mind. A message that doubles belief in the chances of a particular event contains, by definition, 1 bit of information.

On the other hand, what if Macmillan really doesn't expect the launch? If asked, he might have put the chances at 1 in 1,000. With Kennedy's message, the launch suddenly became a certainty.

The number of bits of information in a message is defined by a mathematical formula using logarithms:

$$\textit{number of bits} = \log_2\left(\frac{1}{\textit{message probability}}\right)$$

But you can get the same result without using logarithms: you simply count how many times you would have to double the original probability to get to 100 percent certainty. Each doubling corresponds to an additional bit of information in the message.

For example, you might start by believing that the probability of an event is 0.25, or 25 percent. You would have to double that twice to get to 100 percent. So by this measure, a message that changes your belief from 25 percent to 100 percent has 2 bits of information, and in this strange world it is quite possible to have partial doublings that correspond to a fraction of 1 bit of information.

In the case of the nuclear strike, before he got the message from JFK, Macmillan believed the chances of a missile strike to be 1 in 1,000. Doubling once brings it to 2 in 1,000, doubling again makes it 4 in 1,000, and doubling yet again gets it up to 8 in 1,000. In fact, after doubling fewer than ten times, the chances have reached 1 in 1—it's going to happen, there is no remaining doubt. In the context of Macmillan's initial near-disbelief, the message would contain just under 10 bits of information.

Rather surprisingly, the useful information in a message is not an immutable absolute: it depends on what the recipient believed about the likelihood of the contents of the message before receiving it. The amount of information in the message depends on how much it changes the recipient's beliefs.

Most of the messages we receive don't have that kind of hard edge of belief change and consequent action. Receiving a message from your aunt telling you that she has a cold may modify your actions a little—you might decide not to go to her house for a few days. But basically, it's a weak message.

Those kinds of message obviously don't change the prior beliefs of the recipient and don't have the same kind of quantifiable information value. But their content can be quantified by simply counting how many symbols are needed to convey the message. A symbol can be anything—a letter, a word, a number. You might get a message from a friend and count one thousand words in it: it's a 1,000-word message.

The average English word contains approximately 4.5 alphabetic characters. If you include the spaces between words, the average length is 5.5 characters. So a 1,000-word message is also a 5,500-character message.

Quantifying the information in letters, texts, e-mails, and phone conversations in terms of the number of words or characters they contain is quite admissible, but it's more usual to convert the characters to sequences of binary symbols—1 or 0—and then count the binary symbols. Binary symbols are used so pervasively in computer systems that it makes sense to use the same units everywhere.

For example, in the ASCII code used to represent characters in computers, the letter A is represented by a string of eight binary symbols: 01000001 and a space between words is represented by the string 00100000. Each of those 1s and 0s counts as one bit. An average word contains 5.5 characters, so getting a 1,000-word letter through the mail is the same as getting a $1{,}000 \times 5.5 \times 8 = 42{,}000$-bit letter. There are 8 bits in a byte, so your friend's letter stored on a computer would take up $44{,}000/8 = 5.5$ kilobytes of disk space.

Using the term "bit" to describe both the binary symbol and the unit of information is a little confusing, but it is done because the two are very closely related. In fact, a binary symbol often (but not always) carries 1 bit of information.

Information in sounds and pictures is also counted in

bits. Standard phone connections send 64,000 bits of information each second. The music stored on a CD has about 750 kilobits for each second of playing time, and digital TV uses around 1 million bits per second.

A one-hour digital TV transmission therefore consists of 3,600 million bits. The average English word needs about 44 bits to code it, so the bits used for the TV show could have been used to transmit 3,600 million/44 = 81.8 million words of text. A book might contain 100,000 words, so 81.8 million words is the equivalent of over 850 books—a small library.

Let's return to those "real" messages—ones that change your life, make you take an action, or change your beliefs. What is the most surprising message you have ever received?

Surprise and the information content of a message are nearly synonymous. Surprise is what we feel when our existing expectations are radically changed. The information content of a message is a measure of how much our belief in the probability of an event is changed.

This is probably the most surprising message you are ever going to receive. You've just gone to the kitchen to make a cup of coffee, and there, standing by the fridge, is the Grim Reaper. "I'm still young," you object. But he just crooks his bony finger and beckons you to follow.

This really is quite surprising. Averaged over all ages, your chance of dying today is between 1 in 25,000 and 1 in 40,000. If you're under age forty, it's nearer 1 in 3 million.

So numerically, just how surprising is the Reaper's message? It's equal to the number of bits of information in the message. And that is the number of times you would need to double the probability of your death before you went to the kitchen and got the message that made the probability 100 percent. If you're young (well, younger than forty), you'll find that 21.5 doublings take you from the 1 in 3 million chance of your death to the final certainty of 1 in 1.

So that's it: there are 21.5 bits of information in the Grim Reaper's message of your untimely death.

Here's a less surprising example: "You haven't won the lottery."

The chances of winning the top lottery prize are about 1 in 14 million, or a probability of 0.0000714 percent. If that is the probability of winning, the probability of not winning must be 100 percent−0.0000714 percent = 99.9999286 percent. The fraction of a doubling to get this number up to 100 percent is best worked out on a computer; it comes out at 0.0000001 bits.

That's a very unsurprising result, reflecting the very unsurprising message that you haven't won. It chimes with the saying of one of my politically correct and dry mathematician colleagues:

"The lottery is a tax on the statistically challenged."

SPEAK GEEK

— *A practical application* —

A CD can store the text of 150 Bibles.

If you take a Bible, count the number of words on a typical line, count the number of lines on a page, flip to the last page to find the number of pages, then multiply all those numbers together, you'll arrive at an estimate of the number of words in the Good Book. It should come out somewhere near 800,000.

Including spaces, the average English word is 5.5 characters long. That gives the Bible a character count of 5.5 × 800,000 = 4,400,000. A computer uses 8 bits = 1 byte of binary code for each character, so a text file of the Bible would take up 4.4 MB of storage. That's less than 1 percent of the space available on a CD.

A lot of truth on a very small disk.

10

WAR CHEST

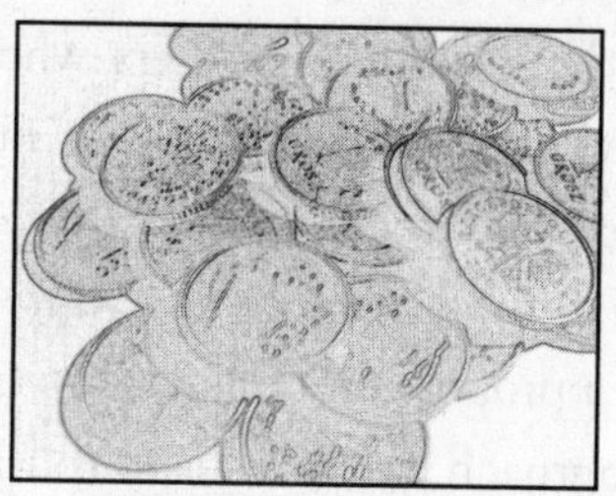

How much tax has to be collected to fight a war?

Drive northwest out of Glasgow for about an hour, and you will come to Gare Loch. The scenery is ruggedly idyllic, sunlight sparkles on the water, and . . . there are thousands of feet of high wire fencing.

This is the Faslane submarine base, home of Britain's nuclear deterrent. A short distance back from the shore, among trees and scrub, you'll find a semipermanent camp of protesters. They're a mixed bunch: the young and rootless, the old and awkward, but also down-to-earth workaday citizens.

Of course, there are many others who don't join peace camps but still believe that possession of nuclear weapons is wrong. Confronted with a democratically elected government that uses taxation to fund weapons and wars, some people have tried to withhold a proportion of due tax in protest. The problems with this approach are manifest. You might not stop at withholding tax used for military purposes: childless couples might refuse to pay for education; the man who has never had a day's illness in his life might claim a rebate from national insurance contributions.

And anyway, just how much money is raised by taxation, and how much of it goes to the big-spending departments of government—defense, education, and social security? I'm going to work it out for my own country, Britain, but you can easily insert your own figures for the United States.

First, we need to know the number of working people in Britain and their average taxable income. Multiplying those two numbers together, and then by the tax percentage, gives an idea of the amount of money that goes to the treasury.

There are 60 million people in Britain, but how many of them are working and paying tax? Let's say that the working population is between 20 and 60 years old: a span of 40 years. A simple estimate of the fraction of the whole population who are of working age is 40 years divided by the average lifespan of, say, 75 years. That's roughly half the population, or 30 million people.

Bear in mind that at any time about 10 percent of the population are unemployed, perhaps another 10 percent are ill, and half the population are women, many of whom take time out during their working lives to rear children. We could realistically reduce our estimate of the working population by 30 percent to account for these factors, which gives an actual working population of 20 million.

This number is quite startling. It tells us that the population of 60 million is financially supported by just 20 million people.

Back to tax revenue. The average income of a working person in Britain is only about £20,000 or $40,000—women still earn much less than men, which brings the average down. After deducting tax allowances, we're left with about £15,000 of taxable income. A 1 percent increase in the basic rate of tax would give the treasury one-hundredth of £15,000 for each person. That's £150 per person, per year. The total revenue would be the number of taxpayers multiplied by £150. That's £3 billion. With £3 billion you could order one item from the following public expenditure menu:

Make benefit payments to 200,000 single-parent families for one year

---

Build one nuclear power station that will generate 1 percent of Britain's electricity for forty years

---

Fund one-sixth of a submarine nuclear deterrent system for twenty years

---

Build and maintain a large teaching hospital for three years

So, how much tax does the government need to raise to fund a war? The Iraq War started on March 19, 2003. By the end of 2006, it was estimated to have cost Britain up to £5 billion, around £2 billion for each year. In the United States these figures are much greater. It can be argued that some of that cost would have been incurred without fighting a war—our soldiers would still have been somewhere—but the operational and ordnance costs would have been much lower. Assuming, in round numbers, that 10,000 British soldiers are involved in the war and that each is paid an average salary of £20,000, the standing cost would be just £0.2 billion. Multiply by two to account for the usual employee overheads and it still only amounts to £0.4 billion—a small fraction of the annual war cost of £2 billion.

Our estimate showed that a 1 percent rise in income tax yields about £3 billion each year, so about 0.5 percent to 1 percent of personal income tax revenue has to be put in the War Chest each year to keep the war running. It's about the same level of expenditure as all the benefits paid to British single-parent families.

But the real cost, the real benefits, and the real negative consequences are incalculable. History is made only once; it's impossible to compare our long-term security, the oil price, and the freedom of Iraqis with what would have been without a war. This is not a good place for Geeks to work, even the nice ones.

SPEAK GEEK

— A practical application —

On average, each dollar in the U.S. economy is used every thirty-six days.

National economies have a fixed amount of money in circulation. The amount is controlled by the government, by minting coins, printing notes, and issuing bonds. It's called the Money Supply.

The money circulates, and the sum of all the country's transactions—shopping, buying a ship, hiring a vet, etc.—in a year is the Gross Domestic Product (GDP). The Money Supply and speed of money circulation are linked by an approximate formula:

$$\text{Money speed} \times \text{Money Supply} = \text{GDP}$$

In rough figures, the U.S. GDP is $13 trillion, and the amount of ready cash in the economy—the so-called Narrow Money Supply—is 1.3 trillion. Using the formula gives a money speed of 10, which means that each dollar is used 10 times per year, once every 36 days.

11

WHEN THE WIND BLOWS

Is a storm more powerful than an atomic bomb?

Katrina, Debby, Grace: they sound so gentle and wholesome. Not at all like destructive hurricanes hurtling across the Caribbean.

Wind is air on the move. Anything that is moving has energy because of its motion, and air is no exception. The energy is usually called kinetic energy, but "motion energy" or "speed energy" would be just as good.

The energy in a hurricane is often compared to the energy of an atom bomb. Could that really be true? Answering the question is easier if the setting is moved to a more familiar location 3,000 miles to the northeast, and instead of a hurricane we just work out the energy in a major cyclone approaching up to Britain from the Atlantic.

If you are British and over 30 or so, you'll remember the Great Storm of October 1987. The wind speed in that storm peaked at around 115 mph. It cut through forests like a scythe. Oaks three hundred years old were toppled like bowling pins. Pine and spruce trees that resisted being uprooted had their trunks snapped like matchsticks. All in all, more than 15 million trees were brought to the ground by the most severe storm to strike Britain in three centuries. The scene in the forest near my Suffolk home resembled a First World War battlefield.

An interesting property of wind is that the force it exerts on an object in its path is proportional to the square of its speed. If you've ever stuck your head out of the side window of a car traveling at 60 mph (definitely not recommended) you'll have an idea of the force of a 60 mph wind. The force of the strongest gusts in the Great Storm would be almost four times as great because their speed was nearly twice that. They would have tumbled you down the street like a rag doll. But was the energy of the storm really on the same scale as the energy of an atomic bomb?

When a vigorous cyclone such as the Great Storm is centered over Britain, a weather map shows the winds revolving counterclockwise in a great wheel whose axle is somewhere over England. The wind funnels along the English Channel, rips up the North Sea, and turns to the west somewhere north of Scotland. Out in the Atlantic, the wind veers to the south before turning back in toward the Channel approaches for its next circuit.

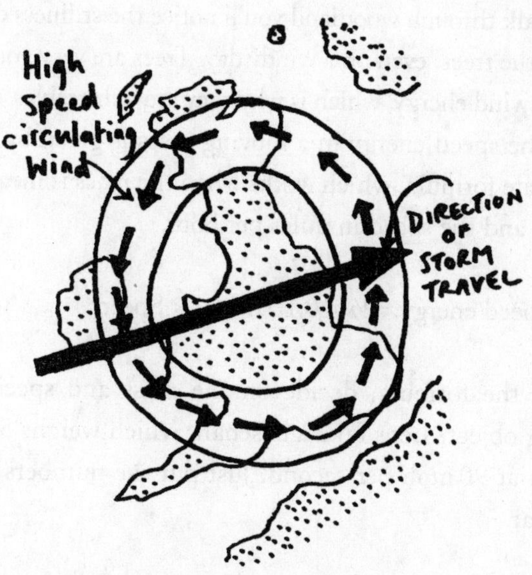

The whole system might be moving northeast at around 20 mph, but the speed of the circulation around the center of the storm can reach 50 mph. It's the circulating air that carries most of the storm's energy.

To estimate the energy of a storm, we need to understand a few basic facts about energy, power, and moving objects. Any moving mass—a car, a rock, air—contains speed energy. This energy can always be changed into heat energy or transferred to some other object by reducing the object's speed. For example, the heat generated in the brakes by bringing a car to a stop from 70 mph would be sufficient to boil a full kettle of water. All that energy was originally tied up in the speed of the car.

When a storm topples chimneys, lifts roofs, or whips up giant waves on the ocean, it does so by giving up some of its speed energy to move the object, and the wind speed is slightly reduced. On a walk through woodland you'll notice the stillness of the air among the trees, even on a windy day. Trees are very good at absorbing wind energy, which is why they are vulnerable.

The speed energy in a moving mass is given by this approximate formula, which works when the mass is measured in pounds and the speed in miles per hour.

$$\text{Speed energy} = \tfrac{1}{2} \times 0.1 \times \text{Mass} \times \text{Speed}^2$$

To use the formula, decide on the mass and speed of the moving object. Let's take a baseball, which weighs 5 ounces, thrown at 90 mph per second. Just put the numbers into the formula:

$$\text{Speed energy} = \tfrac{1}{2} \times 0.1 \times 0.31 \times 90^2 = 126 \text{ joules}$$

That's the amount of energy that would keep a 100-watt bulb burning for more than 1 second.

The joule is a unit of energy named after James Prescott

Joule, a famous English scientist who worked on energy. He was so interested in the conversion of speed energy to heat that on his honeymoon in Switzerland in 1847, he spent much of the time measuring the change in temperature of water after it had fallen down a waterfall.

We can work out the number of joules of energy in a storm in the same way we worked out the energy of the moving baseball. Think of the circulating air in the storm as a car wheel with the mass of the air held in the tire. For a real storm the wheel is perhaps 600 miles in diameter, and the tire is 50 miles deep and as thick as the storm is high, say, 1 mile. The total circumference of the tire is about three times the wheel's diameter, say 2,000 miles, so the volume of air in the tire is $2{,}000 \times 50 \times 1 = 100{,}000$ cubic miles.

Air is surprisingly heavy: 1 cubic yard, the size of one of those dumpy bags of sand you see at the hardware store, weighs more than 2 pounds. One cubic mile of air weighs around 12 billion pounds, and in our storm it's all moving at 50 mph. Putting the numbers into the formula gives:

$$\text{Speed energy} = \tfrac{1}{2} \times 0.1 \times 100{,}000 \times 12{,}000{,}000{,}000 \times 50 \times 50 \text{ joules}$$

That works out to 150 million billion joules, or 150 million gigajoules.

Is that more or less than the energy from the atomic bomb dropped on Hiroshima? That bomb was rated at 22 kilotons, tiny compared with a nuclear bomb. A nuclear bomb can be up to 50 megatons. That's 2,000 times the power of the Hiroshima bomb.

The number of kilotons is how much TNT would have to be used in a conventional bomb to get the same explosive power. The energy of 1 ton of TNT is reckoned to release about 4 gigajoules when it explodes. So, the 15-kiloton Hiroshima bomb had an explosive energy of 63,000 gigajoules. A big winter storm with winds of 50 mph going all the way around Britain has a total speed energy of 150 million gigajoules: about 2,500 times the energy of an atomic bomb, or about the same energy as a big nuclear bomb.

But the energy of the storm is spread throughout a volume of atmosphere of up to 100,000 cubic miles. The atomic bomb that exploded over Hiroshima released its energy into a volume over the city of no more than 100 cubic miles, perhaps one-thousandth of the storm's volume. So the energy per cubic mile released by the bomb would be similar to the energy per cubic mile of the storm.

Much more important, the energy from the bomb was released in a fraction of a second, whereas the storm's energy is dissipated over several hours—a time factor of 10,000 or more. That would make the power of the bomb experienced by the people of Hiroshima 10,000 times greater than the storm. If you stood outside during the Great Storm of 1987, you'd know that the wind's force was dangerous and frightening. The power of the Hiroshima bomb would have been tens of thousands of times greater. Storms cause damage, but they don't melt concrete or vaporize steel.

SPEAK GEEK

— A practical application —

The speed of the gases in a fart can reach 20 mph.

The average speed of gas escaping from an opening can be calculated from the volume flow rate of the gas and the size of the opening. The volume flow rate is measured in liters per second and can be estimated from the volume of gas and the time in which it takes to be expelled.

Volumes are easily measured using an inverted drinking glass when you are in the bathroom. Times are measured by counting seconds. Divide measured volume by time to get the flow rate. Then guess the cross-sectional area of the opening. Assuming a volume of 0.1 liters, an escape time of 1 second, and an opening area of 0.125 square centimeters; the average escape speed works out at 8 meters per second, or about 20 mph. Fortunately for everyone else, the speed drops rapidly as the gas moves away from the source.

If that sounds impressive, it's nothing compared with the speed at which tiny droplets are expelled when you sneeze. They can travel at 157 mph—faster than the fastest scientifically timed tennis serve.

SPEAK GEEK

[illegible]

The speed of the gas as it leaves [illegible]

12

YOU'RE RUBBISH

How many dumpsters could you fill with a year's worth of American trash?

Here's an ideal opportunity for honing geeky behavior. All you need to do is hang out on your street and count trash bags on collection day. On the other hand, if you're a bit shy and don't want to get yourself a reputation, you can do this calculation based on how much garbage you personally put out for collection each week. But it will be more accurate if you lurk on the street.

You'll probably find that there are one or two bags for each house. Houses with just a couple of people might put out just one bag. Houses with a family of four might put out three bags. You need to work out the average number of bags per person per week. Then multiply that number first by the country's population and then by 52 to get the total number of bags of trash thrown out in the United States each year. Then we'll do some calculation on the volume of a bag to see how many would fit into a dumpster.

I'm talking here about the standard type of dumpster that seems to be parked on just about every street: the ones that get filled with building detritus as we periodically replace one set of kitchen units with a shinier model, or demolish an internal wall to knock two rooms into one.

Your discreet observation of rubbish will probably give you a figure of half a bag per person per week. The population of the United States is about 300 million, so that's 300 × 1/2 = 150 million bags being put out each week. To get the number of bags per year, just multiply bags per week by weeks per year. That gives us 150 × 52 = 7,800 million bags per year—call it 8,000 million bags per year.

How much do they weigh? Try lifting a few bags and estimating the weight. It might be helpful to think in terms of the weight of one-liter cartons of milk or juice. These weigh about two pounds. Does the rubbish bag weigh more or less than a shopping bag containing ten cartons of juice? Alternatively, put a trash bag on your bathroom scales.

You'll find that a typical bag of trash weighs about 20 pounds. So the annual weight of household rubbish is about 8,000 million multiplied by 20. That's 160,000 million pounds. There are 2,000 pounds in a ton, so divide this figure by 2,000 to find the weight in tons, and 80 million tons is what we get.

So how many dumpsters can we fill with all this stuff? We need a figure for the volume of the dumpster and the volume of trash in each bag. We'll assume that the trash is compacted before being thrown into dumpsters.

The types of dumpster you can rent vary in size from "mini-dumpsters" with a capacity of around 4 cubic yards to the big ones whose capacity is around 10 cubic yards. We'll go with medium-sized ones with a capacity of 8 cubic yards.

As for the volume of the trash, you can get an idea by jumping up and down on a bag to compact it thoroughly. I suggest you restrict this experiment to your own trash bags and conduct it around the back of your house to avoid getting juvenile probation. I've tried this and found that, with enough jumping, the trash will squash down to about 1.5 cubic feet. That's about 1/20th of a cubic yard.

That's the compacted volume of each of the 8,000 million bags put out each year. The total is 8,000 million × 1/20 = 400 million cubic yards.

How much rubbish will fit into one dumpster of capacity 8 cubic yards? Just divide the total trash volume of 400 million by 8 to get the answer: 50 million. Each dumpster is about 15 feet long, so if they were parked along your street bow-to-stern, they would form a line 15 × 50 million = 750 million feet long. That's 140,000 miles: the line would circle the globe nearly six times.

SPEAK GEEK

— A practical application —

The average American generates ten times his weight in trash each year.

Different countries have different trash habits. Americans create an average of five pounds per day. But, predictably, trash and wealth go hand in hand. American citizens create almost twice as much trash as their European cousins, whereas Russians throw away only a tenth as much.

The anomaly is China. According to one source, the country has only one-fifth of the U.S. per capita GDP, but each Chinese person creates almost half as much trash as an American.

13

THE CUNNING FOX

What are the best words to use in a personal ad?

Recently my niece took me aside to explain speed dating. "There are about thirty men and women in the room," she said. "The women sit down and then the men come around and try to chat you up for three minutes." But here's a strange thing: "Most of the women look pretty average—not ugly, not beautiful. But the men are much more physically varied than in normal life. They range from Mr. Adonis to Mr. Ugly."

At the end of a session the players fill out a card, putting a check mark beside potential matches. Comparing notes with her friends, my niece noticed another gender difference. Women are selective: there are just two or three marks on their cards. Men, on the other hand, play out their biological imperative on paper. They mark more than half the women as potential matches.

A similar ritual is performed every day in the personal ad pages of magazines and newspapers. How about this for a direct, no-nonsense approach?

Cunning old fox seeks sly young vixen for fun nights out and maybe more.

I probably need some advice about this, but if you were looking for a partner, how would you sum up your personality in fewer than a dozen words?

Looking through these ads reveals many of the words and phrases that people use to describe themselves and who they "wltm" (would like to meet). You'll find "fun-loving," "vivacious," and "must have good sense of humor." But what do these lonely hearts really mean?

What exactly are you after when you say that your ideal

partner should have a good sense of humor? Do you mean wittiness, or are you really looking for someone who will tolerate your insensitivity? Or maybe you are just so disorganized that you know it will take the patience of a saint to live with you even for one day.

Take the oft-occurring self-description of "vivacious." Dictionaries define it as "brisk" or "lively." Those words could easily be applied to both men and women, but when did you last see a personal ad in which a man described himself as vivacious? It works both ways: not many women describe themselves as "intelligent and solvent."

There are lots of men who are vivacious in the sense of being brisk and lively, and there are most likely just as many intelligent and solvent women. When applied to personality, however, we all know that there is much more behind those words than their literal meanings. The phrases "vivacious young woman" and "intelligent, solvent man" instantly conjure up images of complete multidimensional people who surely have many other desirable personality attributes beyond the terse descriptions. Try a word-association experiment on yourself. You might come up with something like this.

Solvent

professional

intelligent

tall

responsible

wealthy

dignified

mature

Vivacious
feminine
funny
attractive
slim
talkative
petite
fit

It looks as though personality can be boiled down to just a few well-chosen words, each carrying a complex picture of associations that we infer from our life experience.

With this in mind, you might advise the old fox to add a sprinkling of key words to his ad, like this:

Intelligent, solvent, bushy-tailed old fox seeks vivacious young vixen with good sense of humor for fun nights out and maybe more.

This probably needs a bit more work before the ad is sent off. In fact, the old fox would be well advised to look into a rigorous mathematical method of quantifying his own and his desired mate's personality. He'll then be able to pen the perfect personal ad.

Here's how.

Mathematical techniques have been used to quantify personality for decades. The search for the meaning of personality is rooted in academic research: psychologists wanted to know whether we all had the same basic ingredients in our personalities. A parallel and still debated question was the extent to which personality is governed by genes and by upbringing.

In the late 1940s, the psychologist Raymond Cattell developed a mathematical model of personality. He identified sixteen basic ingredients, such as liveliness, emotional stability, and warmth. Each component has a strength that is given a numerical value.

Cattell arrived at his list of personality ingredients, or *factors,* via a mathematical technique called factor analysis in which the factors are inferred from a statistical analysis of responses to a questionnaire. Each of the questions probes different aspects of personality and is assigned a score.

It's assumed that each person's score on a particular type of question will be governed by the strength of each of the sixteen factors in their personality. For example, the score on a question about how much you like parties might be:

$$(2 \times \text{warmth}) + (8 \times \text{liveliness}) + (1 \times \text{emotional stability})$$

The score on another question about how upset you would feel if a friend got sick might be:

$$(6 \times \text{sensitivity}) + (1 \times \text{anxiety}) - (5 \times \text{emotional stability})$$

In general, the score on a particular question will be a combination of all sixteen factors, but often only three or four will have any significant strength, so the others can be ignored.

If a lot of people fill in a lot of questionnaires, it's possible to check how well the selected factors describe the subjects' personalities. The analysis doesn't actually tell which factors to use, just how well they work. Once a good set of factors has been found, the scores from the questionnaires can be used to make a personality profile.

This is how I think of myself. You may care to introduce yourself in the same way, perhaps by printing it on your business card:

warmth	5	anxiety	4
rationality	8	conservativeness	4
conformity	3	self-reliance	8
boldness	6	tension	7
sensitivity	7	assertiveness	8
trustingness	5	perfectionism	6
abstractedness	3	apprehensiveness	4
privateness	7	emotional stability	6

Doing the math gives us a numerical description of personality, but we want to advise the old fox on the best *words* with which to express his personality in his ad. We don't have the opportunity to subject him to a personality factor analysis, and even if we did, we'd still need to find the right words to express the result. A list of numbers in a personal ad is not going to impress. We need words, and to attract a mate the words must be just right.

A typical ad from a man to a woman looks like this:

Happy philosophical gentleman, six feet tall, passionate and romantic, into art, sports, and the countryside, seeks attractive, young-at-heart slim woman who knows what she wants.

And from a woman to a man:

Attractive, slim, blonde, cosmopolitan, professional woman, 35, seeks witty, wise, and wonderful man to share life with.

A good tip for the old fox would be to use some of the words chosen by other people to describe himself and his desired mate.

Taking the simplest approach, perhaps he should stick to the most commonly used words in *all* the ads. Here they are: the top forty, statistically analyzed from one hundred ads in a national newspaper:

attractive
professional
intelligent
slim
fun
tall
caring
travel
friendship
educated
successful

good sense of humor
sexy lover
fit
feminine
young
wine
sincere
sensitive
genuine
blonde

kind
hair
gentleman
generous
food
elegant
Christian
stunning
strong
solvent
shapely

share	beautiful	romance
loving	arts	sensual
honest		

Here's a possible ad from the old fox that uses words randomly selected from the top forty:

Attractive, intelligent, professional fox seeks tall, slim, caring, educated vixen for friendship, fun nights out, and to share loving and lasting romance.

It sounds over the top, but the statistics show that this is how most personal ads read. Of course, these words are the most frequently used by both "men seeking women" and "women seeking men." Maybe things would be different if we looked at these two categories separately.

Here are the top twenty words that men use to sell themselves to women:

attractive	wine	honest
intelligent	tall	good sense of humor
professional	successful	arts
fun	loving	sincere
travel	lover	share
slim	caring	kind
friendship	sexy	

The first three are the same as in the top forty—it really does seem that the most important thing for dating is to be "attractive," "intelligent," and "professional." But look farther down the list and you'll see that men are likely to introduce words such as "wine," "sexy," and "lover."

Look at how women appeal to men:

attractive	blonde	shapely
professional	sensitive	friendship
intelligent	gentleman	cultured
slim	fun	young
tall	feminine	stunning
educated	travel	solvent
caring	share	

There they are again: "attractive," "professional," "intelligent." But after the big three, the twenty-first-century post-feminist woman typically describes herself in terms of physical attributes: "slim," "blonde," "shapely," and "stunning."

The statistics show the bottom line—men and women want the same main course. The difference is in the side dishes: men would like wine, whereas women go for a bit of care and sensitivity—and solvency.

So far, we've just looked at simple frequencies of words in the ads. But mathematics can do much more. For example, in terms of the ad words, is it possible to pick out distinct types of men or women?

This task requires something called latent semantic

analysis (LSA), a method involving a lot of algebra, which obscures its grounding in common sense. The idea is that underneath all the apparently highly varied word descriptions, there are just a small number of basic concepts.

Look at the top twenty chart for men. The words "intelligent," "professional," and "tall" are all connected in most people's minds. If I tell you that someone is professional, the chances are that you'll also assume that he/she is intelligent and (as studies show) tall. And you'll probably also assume that someone is a he.

Similarly, nobody has any problem in mentally associating the words "slim," "blonde," "feminine," "shapely," and "stunning." What might look at first to be a complex picture with five different attributes is really little more than a one-dimensional description: all those words point to the same mental image.

In fact, the number of broad underlying ideas that we express with language is much less than the number of words available to describe them. Some research suggests that only about three hundred concepts underpin everything we say, hear, and read.

Latent semantic analysis is used to identify the basic concepts used in a collection of texts. Each concept is defined as a combination of words, and each word is assigned a particular numerical strength. Usually, it's found that a concept will have three or four words with high strength and thousands of other words that will be insignificant. The meaning of a particular text—in our case a personal ad—can be succinctly expressed as another weighted combination of just two or three of the basic concepts.

A simplified analysis of real dating ads shows that there are two strong concepts in play. The strengths of the top eight words in each are shown here for the most important concept:

Word	Strength
beneficial	0.59
mutual	0.59
generous	0.31
gentleman	0.30
beautiful	0.29
caring	0.04
fun	0.04
sexy	0.04

And for the next most important concept:

Word	Strength
affection	0.26
mature	0.26
adult	0.26
caring	0.25
intelligent	0.22
fun	0.22
professional	0.16
friendship	0.15

It's easy to make a "map" of dating ads using these two concepts. The value of the most important underlying concept is plotted on the vertical axis, and the second most important on the horizontal axis. Each of the asterisks (man) and circles (woman) on the map represents where a particular dating ad falls in the concept space.

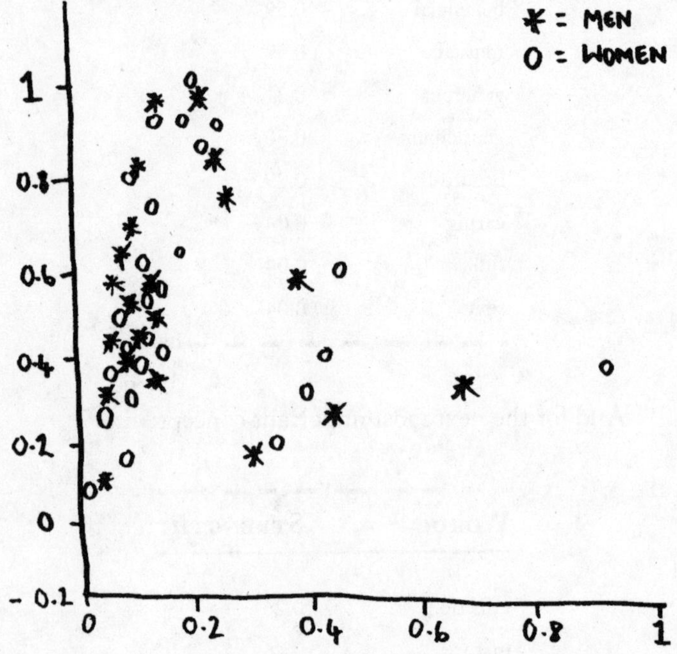

You can see that there are clear clusters of data points. The fox really needs to stick to words from one of these clusters, rather than using words that would place him in an unoccupied part of the concept space. For example, he could use the words from ads that fall somewhere around the middle of the big cluster on the left of the map. Here they are:

arts
attractive
compatible
friendship
intelligent
loving
professional
sociable
tall

We could combine these words like this:

Tall, attractive professional fox, 49, intelligent, sociable and arts-loving, wltm compatible vixen for friendship and relationship.

SPEAK GEEK

— A practical application —

Your personality can be described by just five numbers.

Simple models of personality combine many factors, such as *warmth*, *rationality*, *sensitivity*, *boldness*, and *assertiveness*. But these factors are all related. For example, if you score highly on *warmth*, you are likely to have a similarly high score on *sensitivity*. Likewise, if you score low on *boldness*, you may well get few marks for your *assertiveness*.

This is an inefficient way of describing a personality and gives the impression that variations in personality are much greater than they really are. What is needed is a smaller set of carefully chosen factors that are completely independent of one another.

Fortunately there is a mathematical method called "principal component analysis," which can do exactly that. It can be used to work out sets of abstract personality descriptors that are perfectly independent of one another. Each new descriptor is a subtle combination of the original factors, such as *warmth* and *rationality*, and it turns out that only five of the new descriptors are needed to accurately describe a personality, rather than the sixteen originally proposed by Raymond Cattell.

So, our personalities have only five real components. Instead of sweet, funny, and expansive, with a touch of anxiety, a bit of tension, very apprehensive, warm yet rational, your personality could be described as [1, 3, 1, 6, 4]. What a catch!

14

FLY WHEELS

How many flies would it take to pull a car?

One of the key technologies for personal transportation is a way of storing energy in something that doesn't weigh very much. A gallon of gas will keep your car going for around an hour, with the engine developing an average power of, say, 30 hp (20 kW). The gas effectively stores 20 kilowatt-hours of energy in each gallon.

Gasoline is an amazingly compact way of carrying a lot of energy. A gallon of it weighs only about 6 pounds. A figure of merit for gasoline as a portable fuel might be the amount of useful energy for each pound of fuel—its *energy density.* On this count gasoline is rated at nearly 2 kilowatt-hours per pound (kWh/pound).

I once made an electric bike powered by a pair of lead-acid car batteries carried in baskets. Together they weighed about 50 pounds and could deliver a total energy of only 2 kWh. That's a puny energy density of 0.04 kWh/lb, just one-fiftieth as effective as gasoline.

This, of course, is the curse of electric cars. There is no cheap battery technology that allows anywhere near as much energy to be stored per pound of battery as per pound of gas.

Pedaling the bike, plus its 50 pounds of batteries, back from work on yet another day when the charge had run out, I made a calculation: 2 pounds of sugar digested by a human would give more muscle energy than both batteries combined.

The energy in a 2-pound bag of sugar is almost 4 kWh. Muscles can convert that energy into mechanical work with an efficiency of up to 20 percent. That's an output energy density of 0.4 kWh/lb—not as good as gasoline, but, amazingly, in the same ballpark.

A muscle engine in a car would be a winner: the emissions would be just carbon dioxide, water, and maybe a bit of wind, depending on the food in the engine. And this may not be just the science-fiction dream that it sounds. There is a lot of academic research into so-called molecular motors, the power source of our muscles.

The molecular motors inside your body are built from a particular kind of protein molecule, the shape of which can be distorted, rather like scrunching up a piece of rubber in your fist. The scrunching is done by a chemical called adenosine triphosphate (ATP), which is supplied to the muscle in the bloodstream. ATP is itself synthesized by the sugars from your food.

At a signal from your brain, the chemical fist of the ATP is unclenched—the molecule rearranges its shape into a more relaxed form, providing mechanical force as it goes. There are billions of these molecules acting together, making the muscle pull. Synthetic proteins, which act in a similar way, might provide the basis of future power plants.

Flying insects can do even better. They use similar molecular motors to make the wing beat repetitively with a very high energy efficiency, up to 40 percent. But could harnessing the power of houseflies be the way forward for car travel? If so, how many would you need to pull your car along at, say, a respectable 40 mph?

First you need an idea of how much power a single fly can generate. A very rough estimate of 1 fly-power can be made if you know the fly's weight and how long it takes to rise a certain distance into the air after taking off from a tabletop.

When the fly takes off, it uses energy to lift its weight. Some of that energy is used to accelerate its body and lift it into the air, and some will go into heating its body and warming the air. But as the fly rises, it also increases its energy in another way. This newly acquired energy is called *potential energy*. The higher the fly flies, the more potential energy it gains.

The energy is "potential" because it doesn't come into play until something happens. If the fly stopped moving its wings, its potential energy would change into speed energy as it fell back down and would finally be released as a small amount of heat energy when it crash-landed on the table.

It's a fair approximation to say that the energy expended by the fly in rising to its cruising height is the same as the potential energy it will have once it's there. (It's actually a bit more, but for the purposes of our armchair arithmetic we can let that go.) So if we can work out our fly's potential energy, we can work out the value of 1 fly-power. We just need to divide this energy by the time it takes the fly to reach cruising height, and we have its power in watts.

The formula for potential energy is this:

$$\text{Energy} = g \times \text{Mass} \times \text{Height}$$

The g here is the acceleration due to gravity: the rate at which the speed of a falling object increases. Its value is 9.81 meters per second per second. Call it 10 to make life easier.

But what about mass—what does a fly weigh? Think of something of a similar weight that you can conveniently measure. For example, suppose that a housefly weighs about the same as a grain of rice. It could be half as much, or twice as much, but it is unlikely to be either a tenth, or ten times as much. A rice grain will do nicely.

Count how many grains of rice you can scoop up with a teaspoon, and put ten teaspoonfuls of rice onto the kitchen scales. Divide the weight by 10, and then by the number of grains per teaspoonful. You now have the weight of one rice grain—and the approximate weight of a fly. It'll come out at about 50 mg. That's fifty-thousandths of a gram, which is the same as fifty-millionths of a kilogram.

Next, have a look at some flies as they take off. How long do they take to rise to their cruising height? That's a tricky one.

You can make several rough timings and average them to get a more accurate final value. But even one timing is hard, because the time is so short. One way is to hold a ticking clock against your ear as you watch the flies rise. Older clocks have a tick-tock time of 1 second, but the ticks on some modern clocks are much more rapid. Count the number of ticks per second from your clock to calibrate your "audio timer."

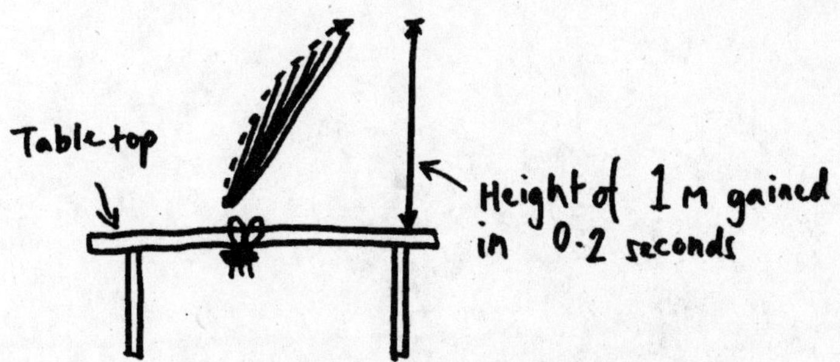

My clock does five ticks per second, and the flies rise about 1 meter in the time of one tick—that's one-fifth of a second. Now the fly-power can be calculated. So, to put into the formula above we have the values 10 (for *g*), 0.00005 kg and 1 meter:

$$\text{Energy} = 10 \times 0.00005 \times 1 = 0.0005 \text{ joules}$$

Power is energy per second, and our fly has taken 0.2 seconds to rise 1 meter, so its power is 0.0005 divided by 0.2. That makes 0.0001 watts—one-tenth of a milliwatt. That's enough to make an LED glow dimly in the dark. A small battery-powered torch gives out about 1 watt—that's 10,000 fly-power. Flies aren't very bright.

A typical car engine running on the flat at 40 mph will be generating around 20,000 watts, or 200 million fly-power. So 200 million flies, attached by silken threads to the front of your car and suitably trained, could pull it along at up to 40 mph. Whether that's a green alternative depends on the flies.

15

BUS STOP

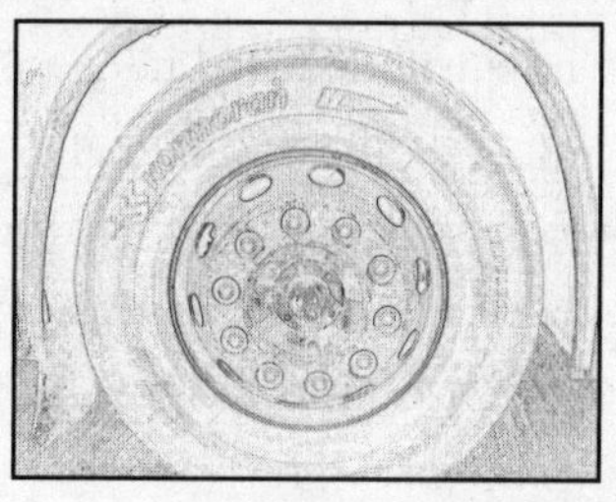

Can you tell the weight of a bus just by looking?

In Sir Arthur Conan Doyle's tale of the "Crooked Man," Watson is visited late in the evening by Sherlock Holmes. Their conversation goes like this:

> "I see that you are professionally rather busy just now," said he [Holmes], glancing very keenly across at me [Watson].
>
> "Yes, I've had a busy day," I answered. "It may seem very foolish in your eyes," I added, "but really I don't know how you deduced it."

There is some knowledge that can be deduced just by standing, quietly observing, and applying logic to your observations. Here's one such example: stand at a bus stop until a bus arrives and pulls up just in front of you.

Look closely at the tire on the wheel in front of you, and you'll see that it has writing molded into the rubber. It might say: "Inflation pressure 56 pounds per square inch." A brief moment of silent deduction and you turn to the driver and say, "I see this bus weighs just over nine tons."

"Yes, that's correct," says the driver. "It may seem very foolish in your eyes," he adds, "but really I don't know how you deduced it."

"Elementary," you reply. "Like this."

Like how? Well, first of all, common sense tells us that the pressure in each tire is what stops them from being flattened by the weight of the bus. A heavy bus is going to need a high tire pressure to stop the tire from going flat, whereas a light bus will need a lower tire pressure. There's obviously a connection between the tire pressure and the weight of the bus. Find the connection, and you will know the weight of the bus.

The tire on this bus is pumped up to a pressure of 56 pounds per square inch. Technically, the tire needs a pressure of 4 bars—the bar being a unit of pressure equal to the average pressure that the air around us exerts on every object—also known as atmospheric pressure. It's caused by the weight of the air above the ground, all the way up to the edge of space.

Atmospheric pressure is 14 pounds per square inch. That means that the atmosphere puts a weight of about 14 pounds on every square inch of the earth's surface, so air is obviously quite heavy stuff.

Thankfully, we can't feel its weight because it pushes evenly all over our bodies on the outside and from the inside. You will feel it if you hold your breath next time you get into a high-speed elevator. As the elevator goes up, the external air pressure will decrease, but with your mouth closed your internal pressure will remain at the ground-floor value. The air will try to burst out through your lungs until you let your breath out through your mouth (the effect is noticeable but not dangerous).

Have a closer look at the bus. Each tire is flattened where it touches the road. The flattened bit is about 12 inches long and 8 inches wide—the width of the tire. No other part of the bus is touching the road, so logically it must be just that bit of tire that supports the entire weight of the bus. Its area is $12 \times 8 = 96$ square inches.

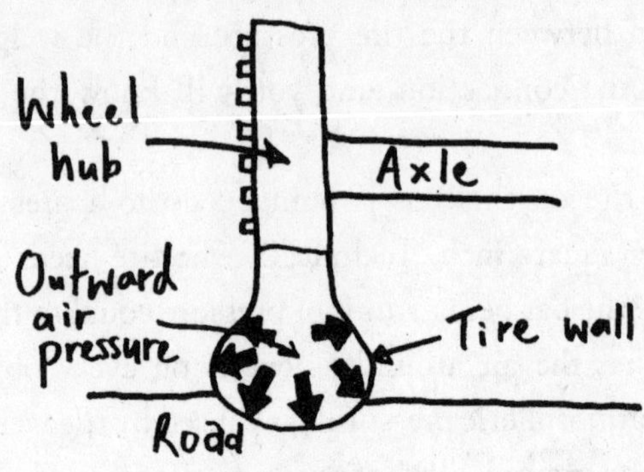

The tire is not flattened to contact more area of road than 96 square inches because of the pressure of the air inside it. In fact, the pressure on the inside of the part of the tire touching the road must be just enough to support the fraction of the weight of the bus that is pushing down on that wheel.

"You see, it really is elementary," you say to the driver, but he still looks puzzled.

"All right, I'll spell it out for you," you tell him.

Each square inch of the flattened area of the tire is being pushed outward by the air pressure inside the tire. That's 56 pounds per square inch, so the total outward force is 56 pounds times the number of square inches of the flattened part of the tire.

You have already estimated the flattened area to be 96 square inches, so the total outward force through the rubber in contact with the road is 96 × 56 pounds. That makes . . . well, this is just an estimate, so make the arithmetic easy by calling it 100 × 50 pounds. That's 5,000 pounds.

Finally, look at the rest of the bus. It has two wheels at the front and a pair of wheels at the back, four wheels in all (not unusual). You have estimated that each wheel carries 5,000 pounds weight. So the total weight of the bus must be about $4 \times 5,000$, or 20,000 pounds.

There are 2,000 pounds in a U.S. ton, so the weight of the bus in tons is the number of times that 2,000 divides into 20,000: about 10 tons.

"Excellent!" cries the driver.

"Elementary," you say. "It is one of those instances where the reasoner can produce an effect that seems remarkable to his neighbor, because the latter has missed the one little point, which is the basis of the deduction."

Reveling in your cleverness, you look up to board the bus, but it's gone.

"Jerk," mutters the driver as he pulls away.

SPEAK GEEK

— A practical application —

AN 8-INCH DIAMETER RUBBER SUCTION CUP COULD LIFT FOUR MEN.

At construction sites you can sometimes see rubber suction cups connected to handles being used to lift heavy plate glass windows. They stay stuck to the glass by the pressure of the atmosphere. The air between the glass and the cup is squeezed out upon contact, so there is no air pressure pushing back on the glass side of the cup. But the back side continues to be pushed toward the glass by the pressure of the atmosphere, so there is an overall force pushing the suction cup onto the glass.

A big 8-inch cup has an area of about 50 square inches, and each square inch is pushed by atmospheric pressure with a force of 14 pounds. That's a total force of 900 pounds, the same as four heavy men. Attach glass plates to the heads of the men and you could lift them all with one cup. You could then hang them as a chandelier in your front room.

16

STIRRING AIRS

How much power does a wind turbine generate?

Where do you stand in the Great Energy Debate? Do you know how much electricity is consumed in the United States each day, or how much comes from coal, renewables, or nuclear generation?

If you think that we should dramatically reduce our energy consumption, are you willing to forgo some of the comforts of modern life? Maybe you believe that we can get our energy from solar, wind, or tidal sources, but will you put up with an erratic, weather-dependent supply and a disfigured landscape? You might be in favor of more nuclear power generation, but in whose backyard are you planning to bury the radioactive waste?

I have a confession. When I was a young man I proudly displayed A NUCLEAR POWER—NO THANKS stickers on the back of my car; I think I liked the artwork. I was genuinely concerned about the problem of waste disposal, but there was a strong element of going along with the herd.

In truth, back then I didn't know too much about what makes life go around, either physically or socially. But of course, without working things out for yourself you can't really engage with debates on the big issues as an independent and informed participant.

Do you know, for example, how much energy is generated by one of those wind turbines that now dot the landscape? And even if you do, do you know how much it contributes to our national energy requirement or how its output compares with the power from a conventional power station? As it happens, calculating an upper limit on the amount of power available from the wind is easier than you might think.

The job of a wind turbine is to extract energy from the

moving mass of air that is the wind. The turbine slows the air, taking some of the air's speed energy, which it converts into electricity.

To make things simple, imagine the turbine placed inside a tube of width, height, and length all equaling 1 meter, and place the turbine at the back end of the tube, with the opposite end pointing into the wind.

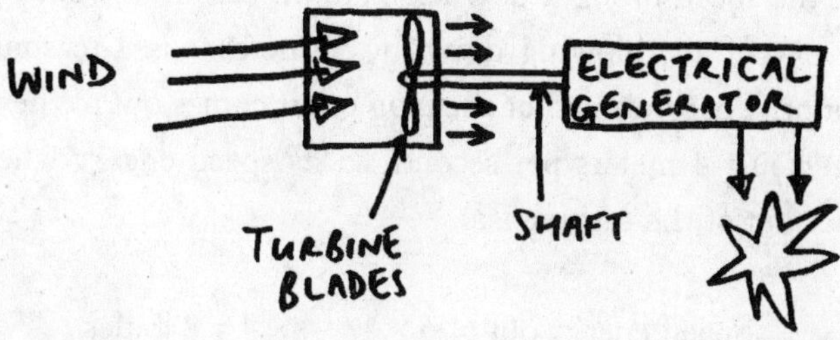

Because the air is moving through the tube, it has speed energy. The amount of this energy is given by the formula we've met before, but to make life easier we're going to use it with mass measured in kg and speed in meters per second. That simplifies the formula to:

$$\text{Speed energy} = \tfrac{1}{2} \times \text{Mass} \times \text{Speed}^2$$

At any instant the tube contains a "plug" of air 1 meter long by 1 meter wide by 1 meter high. The volume of this plug is $1 \times 1 \times 1 = 1$ cubic meter. Now, 1 cubic meter of air has a weight (really, we should call it mass so as not to annoy any physicists) of about 1 kg. That's about 2.2 pounds.

A stiff breeze on a hillside might equate to a wind speed

of 18 mph. That's about 8 meters/second. Putting these numbers into the formula gives the speed energy of the 1 kg plug of air as it enters the tube as:

$$\text{Speed energy in} = \tfrac{1}{2} \times 1 \times 8 \times 8 = 32 \text{ joules}$$

As the wind blows, the plug of air passes through the turbine at the back of the tube, and is slowed down. We can't have the speed brought down to zero because that would stop the flow of air. Slowing down by 50 percent is a reasonable amount. That means that the plug of air comes out of the turbine at just 4 meters per second, so its speed energy when it comes out of the turbine is:

$$\text{Speed energy out} = \tfrac{1}{2} \times 1 \times 4 \times 4 = 8 \text{ joules}$$

The difference between the speed energy of the plug going into the turbine and when it comes out is $32-8=24$ joules. That energy went somewhere, and if the design is good, most of it will be making the turbine turn the electrical generator. So we've got 24 joules from this single plug of air in the tube.

Now, that plug of air is continually being replaced by new air because the wind is blowing steadily at 8 meters per second. That means that eight new plugs of air go through the tube every second, because each plug is 1 meter long. So the turbine is converting $8\times24=192$ joules of wind energy every second.

Joules per second is called power and is measured in watts. The turbine is generating 192 watts—enough to light a couple of lightbulbs, or run a TV or computer.

It's a long way from powering a whole city, but that's

because we made the turbine small: the area of wind caught by the turbine was just 1 meter wide by 1 meter high. If you doubled the area of the turbine you would get twice the power because the mass of air passing through the turbine would be doubled.

We need an estimate of the area swept by the blades of the kind of turbine you see dotted over hillsides. Multiply the result of 192 watts by this area and it will give an estimate of the power generated by a real turbine. The area swept by real turbine blades is a circle, but to make the calculation simple you could approximate it to a square whose sides are equal to the diameter of the turbine blades. (If you are a stickler for accuracy, you can use the formula for the area of a circle: Area = ¼ × π × Diameter^2.)

You can estimate the length of the blades when you next see a turbine. One way is to estimate the number of man-heights, house-heights, or car-lengths that fit along the blade. This really only works if a man, house, or car is conveniently standing next to the turbine.

Still, you'll probably find that the kind of turbine you see in groups on hillsides has a blade length of about 10 man-heights, say 20 meters. That makes the diameter 40 meters, and using our approximation of the square area swept by the blades we get 40 × 40 = 1,600 square meters.

You can now work out the power generated by a real turbine: it is 1,600 × 192 watts. To make life easier, approximate that as 1,500 × 200. The answer is then 1,500 × 200 = 300,000 watts, or 0.3 MW.

So now you know how much power to expect from a really big wind turbine in a moderate breeze. To join the environmental debate you need to know how many turbines there are in a wind farm, how many farms there are in the country, how often there is no wind, and more. It's also worth knowing that a big nuclear station—of the type designed by Westinghouse in the United States—can generate upwards of 1,200 MW.

That's the same as the estimated power from 4,000 wind turbines—when the wind is actually blowing.

SPEAK GEEK

— A practical application —

To generate all of Britain's electrical power from the wind would require more than 100,000 wind turbines.

A wind turbine can generate anything between 100 kW and 1,500 kW of power, depending on its size and the wind speed.

The average amount of power required to keep the United States plugged in is 460 million kW—the same as around 150 million electric kettles. So if we make the optimistic assumption that each turbine can deliver an average power of 460 kW, a total of about 1 million turbines would be needed.

But if the wind didn't blow, there wouldn't be any power, or any tea.

17

DREAM FLIGHT

What size wings does an angel need to fly?

Years ago I lived in an austere apartment complex of gray concrete flats in Helsinki. Each week we received a copy of the *Guardian*, printed on very thin tissuelike airmail paper. One dull winter Sunday afternoon, we cut our store of *Guardian*s into shapes like orange-peel segments and glued them together to make a balloon just over three feet across. More crumpled *Guardian*s were hung from a string beneath an opening in the bottom of the balloon.

Outside, the crumpled newspaper was ignited, and the balloon filled with hot air and ascended slowly up the side of the building. Then the balloon itself caught fire. Its ascent accelerated until at 100 feet above the ground it became a fireball. There was a smattering of applause from Finns who watched the balloon rise past their kitchen windows before Sunday afternoon resumed its gray trajectory.

You've most likely experienced a flying dream. Soaring through the air, unfettered from gravity, seeing earthbound life from a new and detached perspective. Most people have experienced flying as a passenger on an aircraft, but the idea of taking to the skies ourselves seems deeply seated in our psyches and predates any kind of mechanical flight.

Witches ride brooms. Icarus strapped on wings of wax and crash-landed in the sea due to unforeseen technological failure. Modern-day fairy tales have Harry Potter riding a carbon-fiber broomstick, and children cross continents during the night in the secure embrace of flying snowmen.

And . . . angels manage to fly around with the aid of small white feathery wings.

The first real human flight was made by hot-air balloon

in 1783. The "aerostatic globe," designed and built by the Montgolfier brothers, carried a doctor and a military officer about five miles across Paris. In its day, the technology attracted enormous attention from both the public and the military. Generals immediately saw the potential of flight for both surveillance and offensive action.

The development of hot-air balloon technology in the eighteenth century closely parallels the development of space technology in the twentieth. Both technologies were rapidly driven forward by military needs, and both were first tested on animals.

The Montgolfier brothers' first passengers were a sheep, a duck, and a chicken. All survived the flight. The Soviets launched a dog named Laika into space in 1957. Laika didn't make it—no provision had been made to get her back to Earth.

A hot-air balloon generates lift because the air inside the balloon is less dense than the air around it. It rises into the air for the same reason that a bubble in water rises to the surface. Ignoring the weight of the balloon's material and its payload, all that is needed for the balloon to lift off is for the weight of whatever is inside the balloon to be less than the weight of the surrounding air it displaces. A balloon floats in air just as a ship floats in water.

Hot air weighs less than cold air. The weight of 1 cubic yard of air at a temperature of 50° F is about 2 pounds. If you managed to heat the air in a 1 cubic yard balloon to 86° F its weight would fall by 0.2 pounds. That's a weight change of about 3 ounces, and that's how much a 1 cubic yard balloon could lift. Clearly, it will take a big balloon to lift a man.

Say a balloon needs to lift a load of 500 pounds—that's one heavy man plus his weight again in balloon material. The number of cubic yards needed to get it off the ground is 500 divided by 0.2, which is 2,500. If you were able to paste a cubical balloon together, it's sides would each have to be about 13.5 yards long. That's a cube whose sides are each a bit longer than a London bus (length, 30 feet).

But witches and children in fairy tales don't travel sedately by balloon; Mr. Potter, for example, swoops and soars at fabulous speed on his Nimbus 2000. But ordinary people need wings. And wings for a man tipping the scales at 220 pounds need to be big, much bigger than the tiny wings on a quidditch ball or the white fluffy things that angels in pictures have.

You don't need to be an aeronautical engineer to work out some basic facts about wings. You can use pure logic. For example, the area of a wing is going to affect the amount of lift it can generate. A large wing will give a lot of lift, and a small wing only a little. Common sense tells us that the lift will be proportional to the wing's area.

Then there is the density of the air through which the wing moves. The lift happens because of a complicated force reaction between the wing and the air. If there were no air at all, there would be no lift. But if air were very heavy stuff, the forces between a wing and the air would be big. Overall, it is likely that wing lift is proportional to the density of air.

And then there is the speed. Speed is essential. No speed, no lift; stationary aircraft on the tarmac at O'Hare stay glued to the ground. A fixed-wing aircraft has to accelerate down a runway until its speed is sufficient to produce enough lift for takeoff. For a plane such as a Boeing 737, that might be about 140 mph.

But, does a 737 doing 75 mph down the runway get half the lift of the same plane doing 150 mph? The answer is no—it gets a quarter of the lift. This is quite a general rule: the lift given by a wing is proportional to the square of the speed.

The usual way of explaining how speed affects lift is by considering the different distances that the air has to travel over the top and bottom surfaces of the wing. Wings are usually made so that the top surface is hump-shaped while the bottom surface is more or less straight, so that the distance that air has to travel over the top of the wing is longer than for the air passing beneath. Air passing over the top of the wing has to travel faster and expand to compensate for the greater distance. The air expansion reduces the air pressure above the wing and results in an upward force on the wing. A mathematical analysis of the speed energy in the upper and lower flows of air shows that the pressure difference and the hence the lift force are proportional to the square of the speed.

Real wings can behave quite differently from the mathematical prediction, and commonsense logic is enough to get an intuitive understanding of the relation between speed and lift. Imagine that the wing is a slightly inclined flat sheet pushed forward through the air. Air particles collide with the underside of the sheet, and because each particle has a bit of mass, each collision gives a little upward push to the sheet. This isn't a full picture of the way wings work, but it does describe one of the significant effects that give lift to real wings.

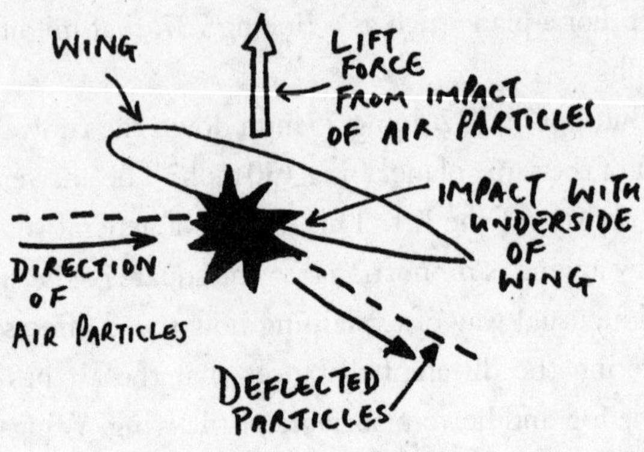

The upward force from each of the air particles is proportional to the speed at which they strike the sheet. The faster the sheet is pushed, the harder the particles strike its underside, and the bigger the lift. But that's not the whole story. The greater the speed, the more frequently particles strike the sheet. So speed affects the lift force twice over: the lift force on a wing is proportional to its speed squared.

This armchair reasoning has shown that the lift is proportional to the density of the air, to the wing's area, and to the speed squared. Here, then, is a formula for lift using square meters and meters per second for the wing area and speed:

$$\text{Lift} = \tfrac{1}{2} \times k \times \text{Area} \times \text{Density} \times \text{Speed}^2$$

You'll see that I've craftily added $1/2 \times k$ in front. Don't be alarmed. The k is what is called the *coefficient of lift*, and it wraps up all the detail about how well the wing is working. The $1/2$ comes from a proper mathematical analysis of a wing.

You can test your formula against a familiar plane like a Boeing 737. I generally do this when the plane is accelerating along the runway; I want to reassure myself that it will indeed take off and not plunge through the traffic on the M4 at Heathrow . . .

The value of k ranges from 0.1 to 1, and being cautious, I assume a value of 0.25 for the 737. Its runway takeoff speed is about 140 mph, which is roughly 60 meters per second, at the point of takeoff. The total wing area is 125 square meters, and the density of air is about 2.5 pounds per cubic meter. Putting these values into the formula gives:

$$\text{Lift} = \tfrac{1}{2} \times 0.25 \times 125 \times 2.5 \times 60 \times 60 = 140{,}625 \text{ lbs}$$

And, thank goodness: the predicted lift of 140,625 lbs is greater than the 100,000-pound weight of this plane. We're going to fly after all!

Let's get back to the size of wings that a man-sized angel would need in order to fly. Angels, of course, have the important task of lifting souls from the dying and transporting them to the place where they are sorted for further relocation to heaven or hell. Estimating the weight of a soul is beyond the scope of this book, so we'll simply assume that the angel has to lift his own weight—about 220 pounds in round numbers (he's a portly angel).

We've seen that the amount of lift depends on the speed of flight. But how fast could an angel fly? To fly as fast as a 737, he would need only tiny wings, but would need enormous power to get the speed.

Let's go for a more reasonable flight speed of 20 mph.

That's the speed of a world-class sprinter in a 100-meter race, about 10 meters per second. We'll give him a coefficient of lift of 0.1 because feathery angel wings will not be very efficient. And to make the calculation easier, we'll approximate the density of air as 2.5 lbs per cubic meter.

Plugging these numbers into our formula, we get:

$$220 = \tfrac{1}{2} \times 0.1 \times \text{Area} \times 2.5 \times 10 \times 10$$

And turning the equation around gives:

$$\text{Area} = \frac{220}{\tfrac{1}{2} \times 0.1 \times 2.5 \times 10 \times 10} = 17.6 \text{ square meters}$$

So each wing is about 9 square meters, perhaps 9 meters long, and 1 meter wide. That's a big pair of wings.

It is possible for a mortal man to power his own flight. In 1979 Brian Allen piloted the *Gossamer Albatross* across the English Channel. The plane was made from ultralight carbon fiber and plastic components, and was powered by furiously pedaling a bicycle-like mechanism connected to a propeller.

The total weight of Brian Allen and the plane was 220 pounds, the same weight we assumed for our angel. The *Albatross*'s wing area was 46 square meters, more than twice the calculated value for the angel, but it was designed to fly slightly slower, at just 18 mph, and therefore needed bigger wings.

But at what speed *should* angels fly? The wingspan of a Sunday school angel looks to be no more than 3 meters; after all, they have to fold away neatly on his back. The width of the wings is around 1 meter, giving a combined wing area of just 3 square meters. With this wing area, the angel would need to achieve a flying speed of 25 meters per second to generate enough lift.

That's 58 mph, a brisk yet dignified speed for the transportation of souls.

SPEAK GEEK

— A practical application —

A cat could be lifted into the air by a cluster of 500 helium-filled party balloons.

Cats, by their nature, don't fly. Helium balloons can help. The balloons float up because they are lighter than the surrounding air. The helium inside a balloon is very low-density, so its weight and the fabric of the balloon we can ignore. The balloon's lifting force is therefore approximately equal to the weight of air displaced by the balloon.

The volume of air displaced by the balloon is of course the volume of the balloon itself. Calculating the volume of air, the weight of which is equal to the cat's weight, gives us the size of balloon(s) needed for feline liftoff.

One cubic foot of air weighs just under 1/10 of a pound, so to lift a fat 10-pound cat, you will need a total balloon volume of 100 cubic feet. You could use one big balloon, or the helium-filled party balloons you can buy in gift shops. These have a volume of about 0.2 cubic feet, so you'd need 500 of them. But do keep an eye out for PETA while you're tying them to the cat.

18

PROCESSING POWER

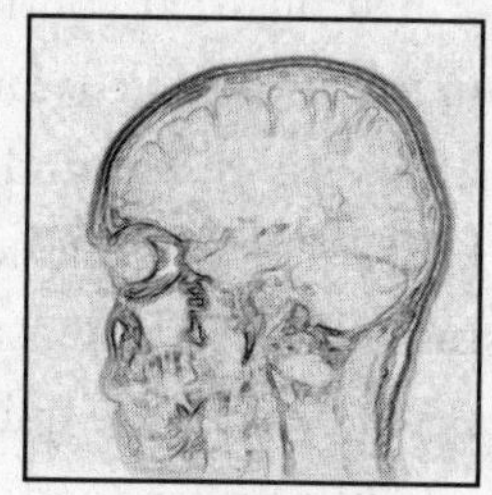

Which is more powerful—your brain or a PC?

An old friend who suffers from Parkinson's disease once showed me its effects. He put two beer mats about nine inches apart on the table in front of us and quickly patted each mat alternately with his affected hand. After ten pats everything seemed fine. After twenty, his arm seemed to have an invisible weight attached to it. After thirty, his hand seemed to be moving shakily through invisible treacle, despite his intense determination.

He explained it to me in layman's terms. In a part of his brain there is an electrochemical representation of his intention to move his hand. But because of the disease, the supply of dopamine, the chemical messenger used to signal his conscious intention to the motor neurons that control his arm, becomes exhausted. The communication link is broken and will start working again only after he's rested and the supply of dopamine is replenished at the neuron synapses.

Parkinson's disease vividly demonstrates the nature of our brains and bodies. While they are quite wonderfully complex, and far from being fully understood, nonetheless they are essentially physical machines.

Mental exhaustion occurs even in alert and healthy brains. Look at the numbers below. Work along the line, adding them up in your head as quickly as possible. Adding the first two takes little effort, adding the next is OK, the fourth is a strain . . . as you go on, it feels more and more difficult to do the next addition.

173, 184, 58, 320, 141, 90, 123, 29

Whether this exhaustion is caused by the temporary depletion of neurotransmitting chemicals somewhere in your brain, or by the accumulation of an inhibiting waste product, functionality has been temporarily reduced. Incidentally, if you truly believe that you can multitask, try doing the addition exercise while listening to the radio.

Here is a crude summary of the physical makeup of your brain. It contains some 30 billion cells called neurons, which can carry signals. Each neuron takes signals from up to 30,000 other neurons, combines them, and transmits the result to other neurons. These signals between neurons are passed as electrical impulses along connection paths to other neurons. When a signal reaches the end of its connection, the electrical impulses cause the release of neurotransmitting chemicals that drift across a gap called the synaptic cleft to another neuron, where they stimulate further signals. Sometimes, the signals may end up being transmitted via the spinal cord to muscles and other organs in the body, but often they remain within the brain, presumably as an internal representation of our conscious thoughts.

Since the late nineteenth century, there have been a succession of descriptions of how the brain functions, each drawing an analogy with the latest complex technology. First the brain was a telephone exchange, switching messages from place to place. In the mid-twentieth century it became a set of regulators, like the thermostat in an oven. Twenty-five years on, the brain metamorphosed into a computer, storing data and doing calculations.

None of these pictures is wholly inaccurate, but each one is a metaphor for some facet of our mental processes. Take the computer–brain analogy. How does your brain compare with a personal computer when it comes to doing arithmetic?

To do a proper comparison, we need a numerical measure of the arithmetic speed of the two systems. For computers, the basic measure of speed is the FLOPS, which stands for floating-point operations per second. A floating-point operation is the name given to one of the basic arithmetic operations done by the processor in a computer, such as multiplying or adding two numbers together.

I just wrote a program on my computer to multiply two numbers together a billion times. The computer took ten seconds to complete the job. So, in each second it did 100 million multiplications, each requiring a floating-point operation. So my computer can do arithmetic at a speed of 100 megaflops, or MFLOPS.

And you: what is your FLOPS rate? Personally, I would be doing very well indeed if I were able to multiply two easy numbers together every second. That's 1 FLOPS, or one hundred-millionth of the speed of my laptop.

Well, Computers 1, Humans 0.

Maybe we could beat the beastly things on memory power. The memory in a computer comes in two types. Random access memory (RAM) is temporary storage space used by the computer while it is doing its processing. Anything stored in RAM is lost when the computer is turned off. The other type is the memory on the hard disk, where files are stored more or less permanently.

Here, a simple comparison between the brain and computer memory falls down because human memory is so integral to the processing part of the brain. In a computer, memory and processing are quite separate.

For this reason, we can't simply count neurons and say

that they are equivalent to the transistor memory cells in a computer's RAM. If that were true, the 30 billion neurons in our brain would be equivalent to about 30 gigabits of RAM. One gigabit is 1 billion bits of information, and there are 8 bits in a byte. On that basis, our brains would have the same order of memory capacity as a PC made in 2007.

And also, human memory is much more complex than units of electronic RAM. For one thing, memory in the brain isn't localized to a few neurons. The information in memory is distributed over thousands, even millions, of different neurons. Destroying modest numbers of these neurons doesn't cause memory to disappear, although the destruction of a very large number of them, for example after a stroke, does impair the overall neural function, temporarily or permanently.

Another big difference between brain and computer memory has to do with definition. You can store photos of your family or music files for your iPod in a computer. The exact pixel description or sound samples stored in those files can be used to re-create, or *remember*, the picture or music indefinitely.

Human memory is different. Most of us can't recall images and sounds in that literal way. What you store in your brain is a higher-level abstraction. You might "remember" the waves, sand, sunbathers, and rocks you saw on the beach on your holiday. But you won't be able to draw a particular wave you saw, or describe it in detail. That's impossible.

Most people don't have a literal memory, but a few have an eidetic, or photographic, memory, which seems to allow them to remember the precise detail of a scene or music. At least, they can correctly answer questions down to a level of detail that suggests they have total recall. That rare ability is often linked to autism and idiot savants.

One of the best-known individuals with such power is Kim Peek, the American on whom the movie *Rain Man* was based. We can get an idea of the potential memory capacity of the human brain by looking at some of Kim's feats.

For example, it is said that he can read a book and remember 98 percent of it word for word. Actually quite a large number of rather normal people can remember books word for word. My mother can quote large extracts from the Bible, and Muslim scholars can do the same with the Koran. The thing about Kim Peek is this: he can repeat the feat for the *twelve thousand* books he has read so far, and presumably for all the books he is going to read in the future. (He can also do it after a single reading—a feat most religious scholars don't attempt.)

So, how many gigabytes of memory are needed to store the contents of twelve thousand books? This is an easy question to answer if you have a computer with a word processor.

Open a document and click on "word count" to get the number of words in the document. Then save the document as plain text. Use the browser to see how many bytes are being used to store the document. I did this for a document on my computer. The document contained 1,897 words and was stored as a file of 11 kB. That means that each word in the document needs an average of 11,000/1,897 = 5.8 bytes per word.

Now estimate the number of words in those twelve thousand books remembered by Mr. Peek. A typical book has 250 pages, and each page might contain 400 words. That's 100,000 words in a book. So in 12,000 books there are

100,000 × 12,000 = 1.2 billion words. Kim managed to remember only 98 percent of the words, but it would still be over 1 billion.

Each word needs an average of 5.8 bytes to store it, giving a total storage requirement of 5.8 billion bytes for 1 billion words. That's 5.84 gigabytes of memory; about twice the RAM in a good PC but just a small fraction of the storage available on a hard drive. So even Kim Peek's prodigious memory is only on the same scale as computer RAM.

You might argue, quite rightly, that Kim knows lots of other things besides the contents of those twelve thousand books. But even if he has a hundred archives similar to his book memory, he would still have a total memory capacity comparable only to that of a large computer hard drive.

I make the score now Computers 2, Humans 0.

Surely humans are better than computers at remembering pictures? Why not give all the benefit of doubt to humans on this one? Imagine that you have eidetic memory and that you have remembered in perfect detail every image projected onto your retina, for every blink of your eyes, ever since you were born.

How many gigabytes of memory of image data would be stored in your brain? To work this out, you need the following numbers: the rate at which you blink, your age, and the number of bytes of information in each image.

I surreptitiously observed a member of my family at breakfast, and they seemed to blink about once every 5–10 seconds. That's 6 blinks per minute, 360 per hour. If a person is awake for 16 hours in the day, that'll be 5,760 blinks each day, around 2 million blinks each year.

By the time someone reaches his fiftieth birthday, he'll

have blinked about 100 million times. And for the purposes of our calculation, each time he blinks he'll need to capture and store the scene in front of his eyes.

If you were to take a photograph with a digital camera, a single image might have a size of 500 kB after it is compressed into the JPEG file format. So the total memory size required for the 100 million images is 100 million multiplied by 500 kB. That's 50,000 GB.

Now, 50,000 GB is a lot more memory than you have on the hard drive of your PC, although with 30–100 PCs you might have it. In a few years' time you may be able to buy a PC with more memory than this. Then you could store every image you have seen each time you blinked from your birth till your death.

That makes the score Computers 3, Humans 0.

Isn't there some way for humans to come out on top in this comparison? After all, who would you prefer as a friend—a computer or a human? Perhaps it's best not to answer that.

The problem is this: we keep playing the computer–human matches on the computer's home ground. Human brains haven't evolved to add and multiply symbolic representations of numbers at high speed. We designed and built computers precisely because we aren't very good at that kind of task. That's why they're called computers, and that's why they win hands down.

But when it comes to picking out a face in a crowd, or recognizing that two pieces of music carry the same emotional message, humans perform much better than computers. So, one last heave for humans to show that we're the masters here. We need a test that humans have evolved to tackle, but which could also be programmed into a computer.

Let's take looking at an image of a face and recognizing its owner. We think we are quite good at this task. Some people seem amazingly good at it. A friend of mine was given the task of showing Queen Elizabeth and Prince Philip around his workplace and introducing them to a hundred or so staff and local worthies. The Queen was delayed, so my friend set off with just the Prince on the great handshaking tour.

When the Queen arrived, the Prince said that he would take her around to repeat the introductions without my friend's help. He did it without making a mistake, and remembered every name.

"How did you do that?" asked my friend incredulously when he got back. "Ties!" he replied. Instead of trying to remember a hundred human faces, he had focused on the different patterns on ties and associated them with the owner's name.

The Prince's task was unusual because it required only a short-term association. But how many faces could you reliably recognize, perhaps when shown a series of photos? For myself, I think it's five hundred to a thousand, and I can probably recall a name, relationship, or some labeling context in a few seconds or less.

Computers are programmed to perform this task, known as pattern recognition, by storing a mathematical representation of each of the known faces. Presented with an unknown face, the computer compares it with each of these representations, and the name associated with the representation that gives the best match is assigned to the unknown face.

One of the commonest methods of matching the stored representations of known faces to the unknown face is to look for periodic variations in brightness across the image. The measurement is done by a mathematical algorithm called the fast Fourier transform, or FFT for short.

The FFT is the single most important mathematical algorithm used in mobile phones, digital TVs, iPods, and other modern electronic devices. Just about anything that processes images or sounds has the FFT algorithm coded into its processor.

The invention of the FFT is usually credited to the computer scientists James Cooley and John Tukey in 1965. The algorithm was actually invented much earlier, but its importance wasn't recognized because there wasn't any technological application for it.

Its real inventor was Cornelius Lanczos, a mathematician and physicist who worked as an assistant to Albert Einstein in the late 1920s. Cornelius did the original work on the FFT in 1940 but never got the credit.

The FFT's great virtue is, as its name implies, its speed. For example, using the FFT to compare two 1 megapixel images requires more than 80 million separate multiplication and addition operations. Previous algorithms would have taken at least a million million operations—quite impractical, even with the fastest computer.

A computer using the FFT to perform the Prince's recognition task would have to do 80 million pieces of arithmetic for each of the 100 candidate faces. That would make 8 billion operations. Your PC will do about 100 MFLOPS: that's 100 million operations each second. So it'll take 8 billion divided by 100 million=80 seconds to process each face.

If you plan to replace the royal visitor with a PC on

wheels, there's going to be quite an embarrassing pause before each handshake as the computer works its way along the line of expectant worthies. Prince Philip can do it in less than 1 second and shake hands at the same time. *We* really are quite impressive. The final score:

Computers 3, Royalty *much more*.

SPEAK GEEK

— *A practical application* —

AN ADULT CAN PERFORM ABOUT 360 ARITHMETIC SUMS PER HOUR.

Many years ago, the British Post Office Savings Bank used hand-operated adding machines to work out the quarterly balance and interest due on each account. The results were then checked by young women who used no more than their heads.

The post office required each of them to process 60 accounts per hour. The record of each account would typically have six different amounts in the old English money of pounds, shillings, and pence (like £3-7*s*-3*d*.). The amounts had to be added and then be multiplied by a percentage such as 2 percent to find the interest due. That's nowhere near the speed of your PC, but it's still pretty impressive.

19

SOUL MATES

Weird coincidence: chance or design?

Raahe, a small town in Finland, was the setting for one of the strangest coincidences in recent years. One day early in 2002 there were two accidents on the main road through the town. The accidents happened within two hours and within one mile of each other. In both accidents a truck struck a man riding a bike. Both accidents were fatal, and both riders killed were seventy-one-year-old men. They were twins, each unaware of the other's movements that morning.

We humans are sensitive to apparent connections between widely separated events within our sphere of perception. Are these freak chance events, or is some strange force at work? It can be straightforward to work out whether what looks like weird coincidence is the result of mere chance or of more mysterious forces. Take those twin brothers in the north of Finland—was that chance or something else?

With this kind of question it's important to start by defining the boundaries of space and time within which you would become aware of a noteworthy event. We live in a world saturated by news coverage—in paper, broadcast, and electronic form—and we would probably hear about an event like the one in Finland if it occurred anywhere in Europe or North America, and we can include much of Asia too. Let's say that's about 3 billion people within the geographical boundaries of our perception.

What do we mean by the time boundaries of the event? It's about how long something remains in our memory. You might recall a strange story several years after the event, and if the event was very dramatic you might keep the memory all your life and recount it to your children and grandchildren. At that point, it could enter folk memory and have an effective lifetime of hundreds of years.

One example of a folk memory is a story about the execution of King Charles I of England. Three and a half centuries later, people still recount the tale of the king wearing two shirts on the scaffold to prevent himself from shivering with cold and leading spectators to think he was shaking with fear.

Strange events like the one that happened at Raahe may stay in our memories until we die, and on average people will be halfway through their lives when they first hear the story. That would put the time boundary of the event at forty years. We'll ignore the fact that some people are forgetful, even of the strangest things.

With the time and space boundaries of our perception fixed, it's possible to start working out the natural chances of the twin death in Finland. If the likelihood of its occurrence by pure chance works out at, say, 50 percent, most people would accept that no other forces were at work. If the odds turn out to be a thousand to one, many would claim the action of stranger forces.

We could start by getting a figure for the number of bikers killed each day on the roads. In Britain there are about 3,000 road deaths per year, and 5 percent of those are cyclists. That's about 150 bikers killed each year out of a population of 60 million, or about one every two days.

The road casualty rate in Britain is quite low compared with other developed countries, and has halved since 1970, even though the amount of traffic has doubled. A more realistic fatal bike accident rate across much of the developed world would be twice the British rate: say 1 death every day for every 60 million people. Put another way, the chance of any individual dying in a bike accident on any particular day is 60 million to 1. (These odds are so high because of the low number of cyclists and their low chance of being killed in an accident.)

The Finnish town of Raahe has a population of less than 30,000, so by simple scaling we would expect a cyclist to be killed only once every 2,000 days. That's about one every five years, and the odds of a cyclist dying on a particular day are 2,000 to 1. Rarely, even in a community of just 30,000 people, two cyclists will be killed on the same day, but the odds are tiny: just 2,000×2,000 to 1. That's 4 million to 1.

The accidents in Raahe were even less likely than that. The two victims weren't just random individuals within the community—they were twin brothers. To refine those odds, we need to know the number of twins in Raahe.

In the developed world, the chance of being a twin is about 1 in 40. So the odds against someone being killed on a bike on a particular day in Raahe *and* being a twin are 40 times the usual odds: 80,000 to 1.

At first sight, the chance of the second twin dying is much smaller because we are looking at the probability of the death of a particular individual; the second victim has to be the first victim's twin brother or sister. If the behavior of the second victim is not linked in any way to the first, those odds are just 60 million to 1—the normal probability for dying in a bike accident.

Without any link between the deaths, the odds of first one twin and then the other twin being killed on the same day are staggeringly small. It's arrived at by multiplying the probabilities of the two separate deaths together. That's a chance of 1 in 80,000 for the first twin, multiplied by 1 in 60,000,000 for the second. The odds are 4,800,000,000,000 to 1 against.

But these phenomenally low probabilities are misleading. There are almost certainly factors linking these two deaths. For example, twin brothers are likely to have similar lifestyles. In particular, if one is a cyclist then it's likely that the other one is too. And once someone gets on a bike, the probability of their being killed in an accident is much higher than average.

We could take an even more extreme position and suppose that twin brothers have identical habits to the point where their every action is identical. Both brothers would leave the house together, ride alongside each other, and incur the same risk of death in the same accident: if one dies then so will the other, because their behavior is identical.

In this extreme scenario the chance of both twins dying at the same time and place is the same as that of a single twin dying, because they would occupy almost the same space at the same time as each other. That brings the odds of both twins dying on the same day on the same road down from 4,800,000,000,000 to 1 to just 80,000 to 1.

The real odds of the twin deaths lie between the two extremes; exactly where, we don't know. One way forward is to be conservative and assume that a single factor links the brothers.

Let's start by saying that because the first brother was a cyclist, the second was bound to be a cyclist. The chance of the second brother being killed in a bike accident is then much higher than for the average person, who is not as likely to ride a bike. In fact, if we assume that about only one in ten people regularly use a bike, the odds against the second brother's death by cycling reduce by a factor of ten, from 60 million to 1 to 6 million to 1.

That brings the odds of the dual death down to a mere

480,000,000,000 to 1. But even after arguing for these lower odds for the dual death, it still looks like the Raahe event is too unlikely to happen by chance. Maybe some other force is at work after all . . .

Except that there's something we haven't yet taken into account. The remote odds apply to all the days within our memory and to all the communities within our sphere of news awareness. Let's take the memory period first.

The odds of the twin cyclists dying on the same day are 480,000,000,000 to 1, but this is where the time boundary becomes important. If you are eighty years old, you'll remember significant events that happened on any day within your eighty-year lifetime. On the other hand, if you have just been born, the number of days you remember is almost zero. On *average,* people will remember events that took place on any day within half of their final lifetime. That's about 40 years, or roughly 15,000 days. The chance of a twin dying on *any one* of those days is 15,000 times greater than the chance for a single day. That brings the odds down to 32 million to 1 against.

Now, 32 million to 1 still sounds rather unlikely, but it applies to just one community with a population of 30,000. We might well remember such an event if it had taken place in any of the communities across our news domain.

The population covered by our news is about 3 billion—nearly half the world's population. Imagine the 3 billion people divided into communities of 30,000, roughly the same as the population of Raahe and its immediate neighborhood. There would be 100,000 communities of that size. In each community, the odds of an accident happening are 32 million to 1, but there are 100,000 communities, so the chance of an accident in *any one* of them are multiplied by 100,000. That makes the odds of us hearing about such an event 320 to 1.

Those odds are still high in relation to commonplace events, and we might still be tempted to start looking for an explanation other than bad luck. "Pure chance might still account for these unusual events." This is all to do with how we register events—our psychology.

This story is about the apparently strange death of twin brothers in cycle accidents, but any similar story would also have made the news. For example, a husband and wife dying in separate bike accidents on the same day, next-door neighbors, father and son, mother and daughter, people with the same birthday, and so on. There is an almost endless list of newsworthy relationships that can be woven around the underlying event of two people dying in cycle accidents on the same day and in the same town.

Let's allow a hundred such relationships into our reckoning—and then call a halt. Each of those variations on the basic story has a similar probability of 1 in 320, but the probability of any one of the stories occurring is 100 times greater. The odds now? Just 3.2 to 1.

We have been quite conservative throughout this calculation, particularly over the linkage between the brothers' behavior, and in limiting the number of story variations to just a hundred. Even so, the chance of this kind of event registering in our awareness is now close to 30 percent.

With a touch of math, the weird becomes likely, and mysterious forces that intervene in human affairs are consigned to the file where the conspiracy theories and alien abduction reports are kept.

SPEAK GEEK

— *A practical application* —

At a party where there are 23 people, the likelihood of any two of them having the same birthday is greater than 50 percent.

Line up the partygoers and ask the first person, a woman, to step forward. There are 365 days in a year, so the chance of her having the same birthday as the second person in the line is 1 in 365.

The chance that she shares a birthday with the third person is of course also 1 in 365. And so on. Each of those individual odds will add up to give the total chance of her sharing a birthday with someone in the line. If there are 23 people, the probability is 23/365=0.063.

Now, repeat the calculation with the second person, but don't count the chance of sharing a birthday with the first person again. The probability this time is 22/365=0.060. Continue the process and add up all the odds. With 23 people the sum will work out at 0.5005, or 50.05 percent.

It's a small world.

20

IDIOT CALCULUS

What can you work out while sitting in a deck chair?

If you've read the novel *The Curious Incident of the Dog in the Night-time* by Mark Haddon, you'll remember how the main character, a young autistic boy, tells how much he dislikes familiar objects in familiar rooms being rearranged. Change makes him feel unsettled, even afraid, until he has remembered the new arrangement down to the minutest detail.

Most people seek comfort and security in familiar surroundings. For many, the comfort lies simply in the existence of a familiar house with the same worn carpets and scuffed furniture, the kettle and teapot just where they always are in the kitchen.

And some of us, I confess, extend this desire for comfort through familiarity into a desire to quantify our surroundings. Take, for example, a child's sand bucket on the beach. Do you know how many grains of sand it holds? Until you know, can you really lie back in your deck chair and relax? I can't.

Imagine a grain of sand placed against the scale of a ruler marked in inches. The grain's diameter might be around 1/32 inch. But the average grain size varies from beach to beach. On a beach of coarse sand it might be 1/16 inch; for fine-grained sand it could be 1/64 inch. The grain size is going to make a very big difference to the number of grains in the bucket.

If the diameter of each grain were 1/32 inch, it would just fit inside a tiny box whose sides were 1/32 inches long. The volume of that box would be $1/32 \times 1/32 \times 1/32 = 1/32{,}768$ cubic inch. You just need to work out how many of these boxes would fit into the bucket by working out its volume. The number of little boxes will be equal to the number of grains.

Buckets are an awkward shape: a truncated cone, like a

cylinder with a taper. You could do some geometry to work out the volume precisely, but an approximation will be good enough. Any error in your estimated bucket volume will be dwarfed by the uncertainty in the size of a sand grain.

Beach buckets are about 8 inches high and have a diameter of around 6 inches at the top and 4 inches at the bottom. You won't make a big error by using the average diameter, and taking the bucket to be a cylinder of diameter 5 inches and height 8 inches.

The volume of a cylinder is its height multiplied by the area of one end. That area—and the area of any circle—is the mathematical constant π multiplied by the square of the radius. But for this estimate it's good enough to do a box approximation again. Imagine that the cylinder is held snugly inside a cardboard box, like those boxes used to pack fancy whiskey bottles.

The height of the box is 8 inches, and its width and depth are both the same as the diameter of the bucket, about 5 inches. The box will have a greater volume than the real bucket

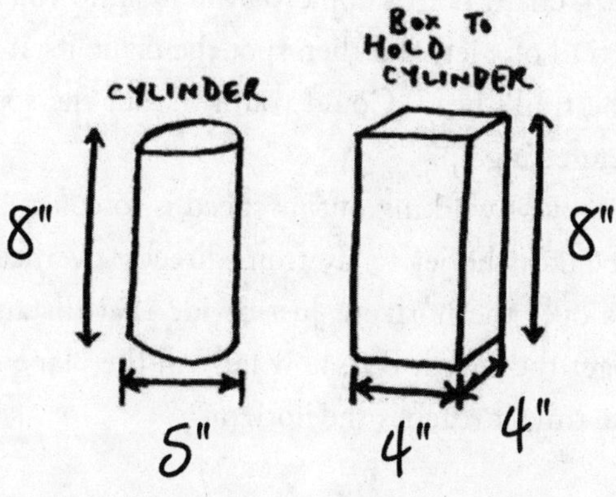

because it has space at the corners that isn't occupied, so you could reduce the effective width and depth of the box to 4 inches to compensate. The volume of the box (and so of the bucket) is about $8 \times 4 \times 4 = 128$ cubic inches.

A bucket of volume 128 cubic inches will hold $128 \times 32{,}768$ or about 4 million sand grains.

What if the sand is coarser than you originally measured, say with 1/16-inch diameter grains? Or, what if the grain shapes are so irregular that they don't pack nice and densely like little balls or boxes?

You could make the estimate again assuming a sand grain size of 1/16 inch. With 1/16-inch grains, the volume of each little box is 1/4096 cubic inches. That'll bring the number of grains in the bucket down to just 0.5 million. Or maybe the grains are smaller, just 1/64 inch in diameter. That would allow 32 million of them to be packed into the bucket.

Well, that feels better. The sand bucket is now fully understood and presents no threat to your peace of mind. But there are many other things that can be calculated as you laze in your deck chair. For example, as you look up you might see the vapor trail of a jet, and then spot the plane itself as a silver sliver at the trail's head. Could you work out the speed of the jet without getting up?

One way of working out its speed is to count how many seconds it takes the jet to fly from directly overhead until it disappears over the horizon. Just divide that distance by the time and get the speed. But, how far will the plane have traveled by the time it reaches the horizon?

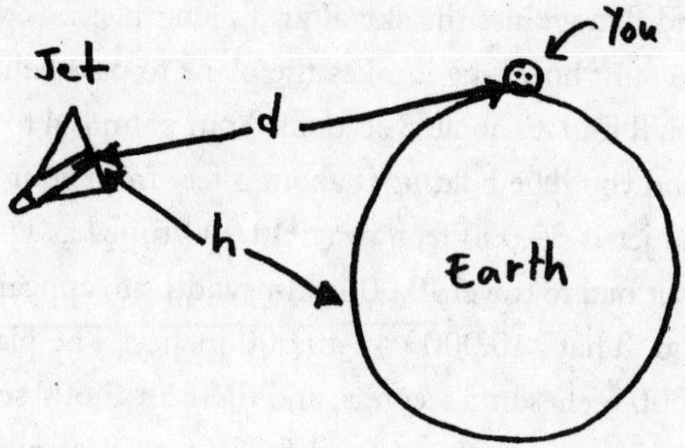

Jets fly at a height of about 30,000 feet, and if you are not near an airport, the chances are that the jet is at cruising height. Working out the distance at which a plane at this height just seems to touch the horizon involves a bit of simple trigonometry and knowledge of earth's diameter. And there's a simple formula. The distance in miles is equal to 1.25 times the square root of the height in feet. The formula works both ways: if your viewing height is h, then it tells the distance of the horizon at sea level:

$$d = 1.25\sqrt{h}$$

The jet's height is 30,000 feet. The square root of 30,000 is 173. So, it will disappear over the horizon when its distance is 1.25 × 173 miles away. That's 216 miles.

But hang on—that's too far, surely? Even if the air were clear and you had eyes like a lynx, it's going to take the jet forty minutes to go 216 miles, and the tide might be lapping round your feet by then. We need another way.

An easier though less accurate plan is to hold the palm of

your hand flat against the sky at arm's length. Look with one eye, and count how long it takes the plane to pass behind your hand. You'll find it's about 5 seconds. Your palm is 4 to 5 inches across, and you'll be holding it about 3 feet from your eye.

The jet is 30,000 feet away: 10,000 times as far as your hand. So it had to travel 10,000 palm-widths to appear to cross your hand. That's $10{,}000 \times 4 = 40{,}000$ inches. The plane traveled 40,000 inches in 5 seconds, and there are 3,600 seconds in an hour, 12 inches in a foot and 5,280 feet in a mile. So, its speed must be $40{,}000/12 \times 5280 \times 3{,}600/5 = 454$ mph.

That's two quite complicated calculations you've managed without recourse to computer, book, or Internet, and if you are anything like me you'll be starting to feel quite rested. But look around—there's more to work out. Like the power station behind my local beach. It hums away 24/7, sending out electric current along cables suspended from giant pylons.

Electric current is a flow of atomic particles called electrons that have almost no mass. So, a question you might ask from your deck chair is what weight of electrons has been delivered from U.S power stations today?

Strictly, the answer is none. Electricity is supplied as an alternating current of electrons. That is to say, the electric field along the wire is reversed sixty times per second (hence the description 60 Hz)—there is no net flow of electrons. (This has the same effect as a continuous one-directional flow of electrons; a so-called direct current.) The electrons move backward and forward a very short distance without, on average, making any overall shift in either direction. However, what you could do is count the number of electrons that just enter and just leave the power station.

Electric current is measured in amperes, or amps for short. One amp (1 A) is one unit of electric charge delivered in one second. A current of 20 A means 20 units of electric charge per second. A tiny packet of electric charge is carried by each electron, and it's the moving electrons that carry the current.

The charge carried by a single electron was first determined at the end of the nineteenth century, but its value was not accurately known until the results of experiments by Robert Millikan were published in 1913 and won him a Nobel Prize. Millikan managed to attach small numbers of electrons to tiny droplets of oil falling through air. The electrostatic attraction of the droplets to electrically charged metal plates allowed him to deduce the charge on a single electron. It worked out at about 1.6e—19 units of charge.

An electric current of 1 A is a flow rate of one unit of electric charge each second; it is carried by 1/1.6e-19 = 6.24e18 electrons per second. A current of 1 A running for 1 hour would transfer 6.24e18 × 3,600 = 2.2e22 electrons.

The daily electricity consumption in the United States is about 11 billion kWh. Electrical power is current multiplied by voltage, so the current is calculated by dividing power by voltage, which in the States is 220 volts. That works out as an average current of 200 million amps for 24 hours every day. The number of electrons that flow in a current of 1 A for 1 hour is 2.2e22, so the total daily electron flow to the nation is 2.2e22 × 200e6 × 24, which comes out at 1.05e32 electrons.

This is a phenomenally large number of electrons. It is 1.05 followed by thirty-two zeros, so the total might be expected to weigh quite a bit. But one electron has a tiny mass, just 9.1e-31 kg. That's a decimal point followed by thirty zeros, then the digits 9 and 1. The total mass of the electrons sent out daily from our power stations to service the electrical energy consumption of the nation for one day is 9.1e-31 multiplied by 1.05e32: about 96 kg or 220 lbs. That's about the weight of a big man.

The tide has come in, with the calming predictable echo of planetary motion. It's time to leave the beach for home. I do hope there's nothing incalculable on the way.

SPEAK GEEK

— A practical application —

A CHILD WALKING ON A STONY BEACH FEELS ONE-THIRD OF THE FOOT PRESSURE FELT BY AN ADULT.

The discomfort you feel walking barefoot on a stony beach is caused by the pressure of the stones on the sole of your foot. The average pressure over the sole of your foot can be calculated by dividing your weight by the area of your sole. Of course, the pressure at the point where a stone sticks into your sole is much greater than the average—that's why it hurts—but the ratio of average to peak pressure will be roughly constant for a given type of pebbles regardless of whether you are an adult or child. Comparing the average pressure for an adult and child will give a measure of the relative discomfort.

Stand on a piece of paper and draw around your foot with a pen. Now, remove said foot and, with a ruler, measure the length and widest part of the outline. Multiply those two numbers together to get the area of a rectangle that would just fit round your foot. Of course, it's more than the real area of your foot, but there's no need to worry about that. Divide that number into your weight, in kilograms, and you'll have a number that is proportional to the average pressure on your foot.

Repeat the operation for a three-year-old; you'll find that your pressure is up to three times greater. Children today have it too soft.

21

THE GHOSTLY PRESENT

Do the dead outnumber the living?

Is the world overpopulated? Lots of people think so. Do you think the United States is overpopulated? How about your town, even your family home? Well, of course, that's different. *We* definitely have a right to be here; so do our family, our friends and "people like us."

Overpopulation is always caused by someone else. It's a fertile field for growing all sorts of conflict, and ostensibly is based on religion, social class, and race. At its root lies the age-old competition for resources that we want to preserve for ourselves and, of course, people like us.

Some ecologists define overpopulation as a state in which the population of an organism is greater than the numbers that can be supported by the resources in the organism's habitat. Grow bacteria in a Petri dish with a limited supply of nutrients and eventually they will stop multiplying because there is nothing left to fuel further division.

In 1789, long before bacterial multiplication was understood, the political economist Thomas Malthus applied the same idea to humans in his *Essay on the Principle of Population*. He believed that in the absence of any control, the human population in any particular locality would increase until their numbers could no longer be supported by the food available from accessible agricultural land. People would then die of starvation, leading to a catastrophic collapse in population.

The food–population system would not necessarily reach a balance: instead it might oscillate through many cycles of population expansion followed by starvation. The unstable fluctuations in population are down to time delays. It takes time for food production to be increased, just as it takes time for infants to grow into full food-consuming adults. The time delays nearly always make the system overshoot the balance point at which the available food just matches the population.

The population of Britain fluctuated around 4 million for many centuries until the Agricultural and Industrial Revolutions (1750–1900). By 1800 it had reached 8 million, and then the growth curve began its current climb. In Britain we're still clinging on to its steep, slippery surface at a population level 7.5 times greater than that of 200 years ago. The reasons for this 52 million increase in population are contentious and complex, but two factors stand out: fossil fuel and technological knowledge.

In the first half of 2008 the price of oil appears to be on an ever upward trajectory towards $150 per barrel. A standard barrel holds 42 U.S. gallons, so the wholesale price is around $3 to $4 per gallon. Add the cost of refining plus distribution, and the price at the pump causes a real pain in our pockets. But think what you can do with a gallon of gas.

Put it in the tank of a 200 hp tractor and you'll be able to plow nonstop for more than half an hour. I watch the twenty-acre field behind my house being plowed every autumn. It takes one man with a big tractor about five hours—that's one acre every fifteen minutes.

Before the Industrial Revolution, a man and horse might plow an acre a day. Even in industrialized Britain, men were routinely using horse-drawn plows up until the Second World War. A relative told me how, as a young farmhand, he pushed his horse to do more than an acre a day. But, he said ruefully, "The horse knew when the acre was done, and wouldn't do more." A modest tractor-drawn plow is around thirty times faster than a horse-drawn one, for a cost of just a few dollars worth of fuel per acre.

Of course, just being able to perform the various processes for growing food thirty times faster doesn't necessarily provide thirty times as much food. There has to be enough land to grow the food, and it has to be fertile land. Until the Agricultural and Industrial Revolutions, soil fertility was maintained by putting animal and even human manure on the fields. Arable farming was also restricted to naturally fertile land.

Since then, artificial fertilizer has made much more land usable and productive. But fertilizers such as ammonium nitrate have to be synthesized chemically using energy from fossil fuels, or dug out of the ground and imported from faraway places, again using fossil fuels, to power the excavation machinery and for transport.

Our present system of food production may not be sustainable; the Malthusian catastrophe could still happen. The king of fossil fuels is coal. Each year on our planet we burn an amount of coal that took a million years to form. The story for oil is similar. At some point they'll run out.

The rapid increase in the world's population over the past couple of centuries, driven by the exploitation of fossil fuels, has led to speculation that right now there are more people alive than have lived before in the whole of human existence. But how true is that—are there really more of us walking the earth's surface than lying six feet under it?

To uncover the truth, it's clear that we need two numbers. First, how many people are alive on the planet at, say, noon today. And second, the total number of people who have ever lived, *and died*, since the apes came down from the trees.

The world's human population today is about 6.6 billion.

Almost none of those people were alive more than 100 years ago. What's more, they are nearly all going to die over the next 100 years. That's at least 6.6 billion deaths over the next century. That's the key to answering the question. We can count all the people who have ever died by going back through time in steps equal to a lifespan. At each date we'll add the population of the world to a count of the dead.

The counting will be easy if we assume that everyone lives for the same number of years, equal to the average lifespan of a human. We're dealing with total numbers across the planet, so there is nothing wrong in using an average number. These days we can look forward to a lifespan of around eighty years. But for much of the world's population, life is shorter. Maybe sixty to seventy years is nearer the mark.

Going back five thousand years, we find from archaeological evidence that our own ancestors lived only into their thirties or forties. So, for the purposes of counting the dead, let's assume that the average lifespan over the whole of human history has been fifty years. And we'll use 2007 as a starting point.

If everyone lived to the age of fifty and then died, we could reason that the entire population of fifty years ago, that's 1957, is now dead. If we knew the world's population in 1957, we could use it as the first contribution to our dead-count. That's just the start. Go back another fifty years to 1907, and suppose that all those people were dead by 1957. Add them to the dead-count and repeat the addition until, say, 50,000 years ago to get a rough figure for the dead.

You may have spotted a small problem with this approach: the number of steps of fifty years to cover 50,000 years is 1,000. Do you really want to add a thousand numbers to get the answer? To avoid anxiety on this score, I can reveal that accounting for deaths in just the last few hundred years is going to be sufficient for a good estimate of the number of people who have ever lived and died. More of that later.

To do the count, you need to know the world's population at each of those fifty-year intervals. You might be able to look it up in a book or on the Internet, but it's more interesting to estimate it yourself. Here's how.

The population is growing. It grows because, on average, each set of parents has more than two children who live to have children of their own.

We are often quoted as having, comically if not accurately, 2.4 children in a family. That would mean, on average, one person becoming 1.2 people after about the period of one generation—say twenty-five years. After another twenty-five years those 1.2 people have children. The number goes up to $1.2 \times 1.2 = 1.44$. That's an increase of 44 percent in the size of the breeding stock over fifty years, about 1 percent per year.

The interesting thing is not the actual numbers, it's the numerical process: more people means more children means more children. The result is that the population increases by a certain factor in a given time period. For example, say the population increases by 44 percent over fifty years. That means that after fifty years the population is 1.44 times larger. After 100 years it would have grown by a factor of $1.44 \times 1.44 = 2.07$, and after each additional fifty-year period this factor will be multiplied by another 1.44.

The birth rate in Britain, for example, is now lower than in most other countries, and our population trend isn't representative of the rest of the world. Fifty years ago, the world's population was about half the current size of 6.6 billion, so on a planet-wide scale it appears that the population seems to roughly double every fifty years. You can use this multiplication factor to write a table of estimates for the population going back through time in fifty-year steps. Every time you go back fifty years, just divide the population by two:

YEAR	ESTIMATED POPULATION
2007	6,600,000,000
1957	3,300,000,000
1907	1,650,000,000
1857	825,000,000
1807	412,500,000
1757	206,250,000
1707	103,125,000
1657	51,562,500
1607	25,781,250
1557	12,890,625
1507	6,445,312

The table is surprising—and *wrong*. It's telling us that five hundred years ago the world's population was just 6.4 million. This goes to show how a simple mathematical model can give completely false results if we push it too far, in this case by trying to extrapolate a long way back into the past.

It is always worth running a sanity check on a numerical estimate. The idea is to compare the estimate with one or two pieces of indisputable evidence (they are often called ground truths). For example, historical records show that in 1500 the population of Britain was about 4 million. Today it is nearer 60 million. That's a ratio of 15 to 1. But according to our table, populations have grown by a factor of a thousand since 1507. Conclusion: the table is wrong.

So the way in which populations grow is much more complex than our simple idea of people breeding like dividing bacteria in a Petri dish of unlimited nutrients.

Instead, we'll have to rely on historical and archaeological records. These show that the world's population was about 5 million in 10,000 BC and grew to between 200 and 400 million by the time Jesus was born. It then rattled around that figure until, between 1700 and 1800, things began to behave much more like our table, with the population doubling every fifty years. In fact, the numbers in the first five rows of the table are about right.

Well, after all that, you can now count the dead and answer the question: are there more people alive today than have ever died? The number of dead is the sum of all the population numbers at fifty-year intervals starting at 1957 and going back to, say, 10,000 BC. The population started at around 5 million at the beginning of this period and finished up at 400 million before the Agricultural and Industrial Revolutions. We could take an average population of 100 million.

It takes 240 steps of fifty years to get back to 10,000 BC. That's 240 lifespans. The number of people who have died in that period is the average population of 100 million multiplied by the number of lifespans. That's 240×100 million $= 24$ billion: about four times the current living population.

So, the answer to the question is that the living *don't* outnumber the dead. On the other hand, our table does show us that the number of people alive today is greater than the number of all the people who lived and died in the past couple of centuries.

SPEAK GEEK

— A practical application —

IF PRESENT POPULATION TRENDS CONTINUE, IN 400 YEARS' TIME THERE WILL BE 1,000 SQUARE FEET OF LAND FOR EACH PERSON.

The total land area of the world is about 60 million square miles. If you divided it into little plots of 1,000 square feet—about the total footprint of a house in a new subdivision—you'd end up with 1.5 trillion of them.

The world's population is currently around 6.6 billion and has been doubling every fifty years or so for the past two centuries. To find how long it would take for the population to grow to 1.5 trillion, count in steps of fifty years and double the population each time until it gets to that figure. It's around eight steps—that's 400 years. There would then be just one plot measuring about 32 by 32 feet for each person.

It gives new meaning to "pressing the flesh."

22

BAD BREATH

Get a sniff of Caesar's last gasp.

In 1948 a graduate student called Clair Patterson began work at the University of Chicago on a way to determine the age of the earth. His method relied on measuring the amount of lead and uranium in old rocks. The trouble was, he found that his rock samples were contaminated with lead—from the atmosphere.

The lead in the air came from lead tetraethyl, a compound, which until recently, was commonly added to gasoline to make it burn more smoothly. The lead added to gasoline worldwide from the 1920s on is still around today in the atmosphere, the oceans, and the ground. It's toxic, but we have no choice but to live with the consequences of our past. It is the ultimate bad breath.

Lead tetraethyl was first used as an additive in 1921 by engineer-turned-chemist Thomas Midgley. But his contribution to long-term environmental damage didn't stop there. In 1928 he developed a chlorofluorocarbon (CFC) as a nontoxic replacement for the ammonia that was used in early refrigerators. It's now accepted that CFCs are the main cause of ozone depletion in the upper atmosphere, as well as being a potent greenhouse gas.

Our atmosphere is not infinite. Some of the things pumped out by various human activities break down harmlessly; others hang around for a long time. The earth is more or less a sealed unit: the only substances that leave the planet in significant quantities are the lightest gases, hydrogen and helium, which tend to boil off into space. Some other substances, such as compounds of lead, are eventually absorbed into rock but only after many millennia. Meanwhile, they've taken up residence in our bodies, plants, oceans, and soil.

Now, breathe in, and hold your breath.

Your lungs now contain a sample of the atmosphere-borne pollutants of our modern industrial age. But that's not all—the couple of pints of air now inside you also includes small quantities of all the stuff that has been pumped out by people, plants, and machines over many centuries.

Take Caesar's dying breath. You probably just inhaled a few of the molecules from the air that passed between his lips as he collapsed on the steps of the Senate and gasped "Et tu, Brute?" You can work out just how many molecules if you know the volume of his lungs, of your lungs, and of the earth's atmosphere.

Of course, we have to assume that Caesar's last breath has been thoroughly mixed into the whole atmosphere and hasn't been permanently absorbed somewhere. The time it takes for the atmosphere to be recycled by dissolving in the oceans and then being released, or being transpired by plant life, is an interesting question in itself—but one we'll have to put to one side.

Start with the volume of air in Caesar's last breath, and the volume of air in earth's atmosphere. Dividing the first number by the second will give the tiny fraction of a liter of Caesar's breath that's now present in each liter of atmosphere.

Next you have to work out how many liters of air are inhaled in one breath. Multiplying that number by the concentration of Caesar's breath in the atmosphere will give the actual volume of his breath you inhale. It's going to be minuscule, but there are a huge number of molecules in even one liter of air, so even in a small volume there will still be a fair number. To get that number, just multiply the volume in liters by the number of molecules per liter. That will give the number of molecules of his dying breath you just breathed in.

It's time to put some actual numbers into this estimate, starting with the volume of one lungful of air. A deep breath amounts to one liter, about the same as a carton of orange juice.

Then there's the volume of earth's atmosphere: that's the area of our planet multiplied by the height of the atmosphere. You could look up the area of earth with Google, or use the formula for the area of a sphere ($4\pi r^2$) to work it out from its radius 4,000 miles. It comes out at about 200 million square miles.

The thickness of the atmosphere is more difficult to pin down because its density and pressure aren't the same all the way up. The air at ground level is squeezed by the weight of all the air on top of it, so it's quite dense. The air at Mount Everest's summit doesn't have so much air above it. As a result its atmosphere never really ends, and the air just gets thinner and thinner. But we need a number for its thickness.

There's a simple way of dealing with this by thinking in terms of an "effective thickness" of the atmosphere: how high it would be if its density all the way up were the same as it is at the surface. Some undergraduate math shows that the effective thickness of the atmosphere is the height at which the pressure of the real atmosphere has dropped to 37 percent of its surface value. That turns out to be about the height of Everest, which is close to six miles.

Multiplying the earth's area by the effective thickness of the atmosphere gives a volume of 1,200 million cubic miles. There are a 4.1 million million liters in a cubic mile, so the volume of the atmosphere in liters is around 5,000,000,000,000,000,000,000. That's 5 followed by 21 zeros, or in our shorthand notation, 5e21.

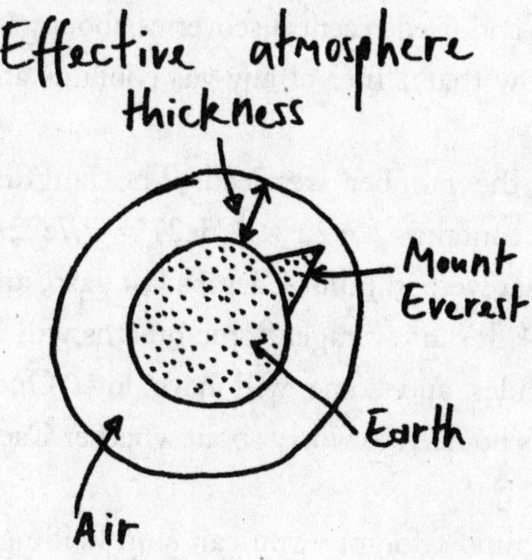

So if Caesar's last breath had a volume of 1 liter, it forms one part in 5e21 of the air we breathe. If your last breath was 1 liter, you just breathed in 1×1/5e21=1/5e21 liters of Caesar's last breath.

All that remains is to estimate how many molecules there are in this tiny volume. It's done using an important number named after the nineteenth-century Italian scientist Amedeo Avogadro.

Avogadro was a professor of physics at the University of Turin. He was also a bit of a political activist and was removed from his post in 1823 so that he could "spend more time on his research"—the nineteenth-century equivalent of "spending more time with one's family."

Avogadro realized that a weight, called the *mole*, of any substance always contains the same number of molecules. In the 1860s the Austrian physicist Josef Loschmidt worked out what Avogadro's number was (Avogadro himself wasn't able to calculate it), and used recent discoveries about gases and molecules to show that 1 liter of any gas contains around 2.7e22 molecules.

That's the number we need. The lungful of air you breathed in contains 2.7e22 × 1/5e21 = 2.7e22/5e21 of the molecules that were in Julius Caesar's last gasp, and *that* works out to be 5.4. It's an average. Some breaths will have none of these molecules and some will have lots. One thing's for sure—there's no need to worry about whether Caesar had halitosis.

Oh, I almost forgot—you can stop holding your breath now. Let it out, and think about all the people who'll be breathing it in a couple of thousand years from now.

SPEAK GEEK

— *A practical application* —

After topping off your quarter-full cup of tea ten times, there would be less than one-millionth of the original milk in the cup.

Many people have the habit of drinking half a cup of tea with milk and then replenishing with tea from the pot without adding more milk. If you drink three-quarters of each cup before topping off, only a quarter of the original milk content remains. So, after ten top-ups the final milk concentration will be reduced by a factor of $1/4 \times 1/4 \times 1/4 \times \ldots \times 1/4 = (1/4)^{10} =$ 1/1,048,576. That's pretty strong tea.

A similar process of successive dilution is used in the preparation of homeopathic medicines. A solution of the supposed therapeutic ingredient is diluted, typically with 99 parts of pure water or alcohol to one part of the original solution. The process doesn't stop there: the dilution is repeated, this time starting with the already diluted ingredient. The process of successive dilution is repeated up to fifteen times before the final solution is administered to the patient.

A calculation similar to the one for working out the strength of your cup of tea gives the concentration of the homeopathic solution. It's $1/100 \times 1/100 \times 1/100 \times \ldots \times 1/100 = (1/100)^{15} = 1e - 30$.

But a 300 ml, about one full cup, of the solution would

contain a total of only about 1e25 molecules. That means that the number of molecules of therapeutic ingredient in the cup would be just $1e25 \times 1e-30 = 1e-5$, or 1/100,000. It's not possible to have a fraction of a molecule: what this means is that there is a chance of only 1 in 100,000 that there will be one of those molecules in the cup of solution you drink. The proponents of homeopathic remedies get around this by suggesting that the water or alcohol used in the dilution process "remembers" the earlier presence of the active ingredient.

Maybe.

23

THE FINAL JUDGMENT

How will your account look when you die?

"And I saw the dead, small and great, stand before God; and the books were opened: and another book was opened, which is *the book* of life: and the dead were judged out of those things which were written in the books, according to their works."

That's the cheery Judgment Day lineup described in the book of Revelation. So, how will the *Book of Your Life* read when it is opened? Is it possible to put some numbers to the sum total of your life?

Of course, there are many things that could be included in the *Book of Your Life*. The credit column of the ledger might list the number of small acts of kindness you performed, the number of children you raised, and the amount of wealth you created. The debit column will record disloyalty to friends, the times you allowed greed to direct your actions, and maybe your personal carbon footprint.

Money is arguably a good measure of your worth to the community as a whole: for a statistician, a Judgment Day starting point would be the average monetary value of a life. Your life really does have a finite monetary value: if life were priceless, you wouldn't travel in a car, plane, boat, or train unless it were 100 percent safe.

A zero-risk plane or train would cost an infinite amount of money, which is another way of saying that it's impossible to make anything completely safe. Typically, doubling the amount of money spent on safety systems will decrease the risk of failure by a small percentage. Doubling the amount again will lead only to a similarly small reduction in risk.

On the other hand, we wouldn't tolerate public transportation on trains and jalopies that had been built for next to nothing. The reason: our lives are worth more than the cost of improving the safety of the train.

So we compromise: we spend a lot of money to make transit systems, industrial plants, and so on reasonably safe, but the amount spent is related to the perceived monetary value of a life. We don't spend too much, and we don't spend too little.

There's another uncomfortable reality: the monetary value of a human life depends on the wealth of the community in which that life exists. In Western countries we spend a lot to make life fairly risk-free. It's a reflection of our wealth and ultimately of the value we place on our lives. But move to a poorer country—Honduras, say—and it's unlikely you'll find the cars, trains, and planes as safe as in the United States.

The value of your life also depends on who's picking up the bill. We demand high safety standards in a nuclear power station: if you lost your life because of a nuclear accident, you would expect your family to receive a hefty financial payout from the operator. If, though, you climb into your car and floor the accelerator, the risk of your dying is hundreds of thousands of times greater than death by nuclear accident. But you have made the choice to risk your own life; it's your life and you can dispose of its value as you please.

So, how much is your life worth? For a rough guide, your life insurance policy will tell you the most that the company will pay out if you die: my policy promises $1.5 million. Or, look in the newspapers to see how much compensation is paid to the families of victims of train wrecks or industrial accidents. All these amounts are in the same ballpark. In fact, companies running public transportation, power stations or industrial plants do quantitative risk assessments on their business and put the value of a human life into the cost–benefit equation. It's reputedly around $6 million.

In the entries in the *Book of Your Life*, the monetary value of your life might be equal to the net profit we made from birth to death. The average annual income for a male working adult is about $68,000. Work for forty years and you'll have made $68,000×40, about $2 million. Yes, most of us will nearly be millionaires by the time we stop working. The trouble is that the money trickles in over your working life rather than arriving as a lump sum. Notice that it's comparable to the payouts made by insurance companies for a death, again indicating that your life has a monetary value that reflects your total earnings.

But, as for any business, your life accounts should show expenses as well as income. As far as the community is concerned, your worth is the net amount by which your birth and life increase the community's wealth. On this basis, the value of your life is the amount you earn minus the amount spent on keeping yourself alive so that you can continue to earn. Those expenses include food, shelter, education, health care, and pension contributions.

We generally pay for food and shelter ourselves. Take shelter: the average price of a house is around $265,000 and is shared by two people, so shelter for one person for life costs around $130,000. That's assuming the house lasts for about one lifetime.

Then there is fuel to keep you warm and for cooking. Add your electricity, gas, and oil bills for a year and multiply by the eighty years of expected life. It'll come out at around $1,000 per person per year, making your lifetime fuel bill $80,000. And of course there is water and sewage disposal: that's another $500 per person each year. Forty thousand dollars for a lifetime should cover it.

Food is amazingly cheap in the developed world. The weekly bill for essential food for one person is about $40, though many of us spend considerably more on all sorts of delicacies that are not really needed. If you live for eighty years, you'll have had 4,000 weeks of essential food: a total bill of $160,000. You might double that figure to $320,000 if you have gourmet tendencies.

The expenses so far total $590,000. Realistically, you'll also be spending money on frivolous luxuries like clothes. That's going to push your expenses up by at least $300,000.

Transportation is very expensive. It's largely used for chores like work and shopping. It's not a frivolous form of consumption, and you might do it by bus, train, plane, or car. If you choose to use a car, perhaps jointly owned with your partner and driving 20,000 miles per year, you'll have an annual expenditure of $4,000 on fuel, around $1,000 on depreciation and maintenance, and $1,000 on tax and insurance: a massive $6,000, or $3,000 per person. That's $240,000 in a lifetime—one and a half times your essential food bill.

And you still haven't paid for education, pension, health, and defense. These items are paid mainly through taxation, and at current rates someone earning $48,000 will pay nearly $10,000 in income tax. Deducted over forty years of your working life, that'll amount to another $600,000 of expense.

So, the financial section of the *Book of Your Life* is going to look something like this:

ITEM	INCOME	EXPENSES
Earned income	$2,000,000	
Food		$320,000
Shelter		$150,000
Heating and cooking fuel		$80,000
Water		$60,000
Clothes, etc.		$330,000
Transportation		$260,000
Health, education, etc.		$600,000
Subtotals	$2,000,000	$1,560,000
Net profit		**$660,000**

This part of your ledger isn't looking too bad. But the wealth you generate over forty years physically decays. It's the Second Law of Thermodynamics applied to your nest egg. The entropy of any closed system increases over time; the fallen log rots away, the bricks in your house weather and turn to dust, the car engine wears out.

All the *real* wealth you will have generated in a lifetime is physical: you can't eat $5 bills. Real wealth is bricks and mortar, machines, and surplus food. Its value has an abstract monetary representation, but wealth is real physical stuff. And it all decays with time.

The value of hi-tech goods vanishes rapidly—computers, for example, are usually discounted to zero over just four years.

The value of a car reduces by about 50 percent every two years. Bricks and mortar are a better investment: the effective life of a domestic house is around fifty years.

At a conservative guess, most of the wealth you created at the beginning of your working life will have been dissipated by the end of your life, while the wealth created at the end of your life will be undiminished. As a result, the difference in the communal wealth at the end of your life and the wealth that would have been if you had never been born is probably no more than half of the $660,000 suggested by the raw numbers.

So, it looks as if the real economic value of your life from cradle to grave is only around $220,000—not much more than $2,000 for each year of your life. But at least the balance is positive: you won't be cast into the Lake of Fire for this bit of the *Book of Your Life*.

But look carefully at the words from the book of Revelation with which we started. It says "according to their *works*"—that's plural. What one does for money is not the only kind of work. Surely selfless piety ought to count for quite a lot in the credit columns. Whose life has the greater worth, Mother Teresa's or Rupert Murdoch's?

Rupert Murdoch has created a business empire that employs tens of thousands of people. Each of those employees is generating wealth because of Murdoch's skill in creating and maintaining the organization. Without News Corp., the average income in Britain might be slightly less than its current level. If it were just $10 less, we could argue that Mr. Murdoch is responsible for creating an annual wealth of $10 multiplied by the working population, say $1 billion. That's a lot more than any ordinary individual is going to create in a lifetime.

Mother Teresa founded and maintained a religious order that has cared for thousands of orphans, the disabled, and sick people. How should we value acts of kindness, which we might crave ourselves but find hard to give: making friendly eye contact with a bum on the street, or giving a word of encouragement to an insecure child?

It might not seem possible to place any monetary value on that kind of work. But kindness, compassion, respect, and care are behavioral traits—traits that hold people together, allow social organization, and ultimately facilitate the economic activity that drives our modern industrial world. They have economic value; without them, News Corp. wouldn't function. They are just difficult to work out.

SPEAK GEEK

— *A practical application* —

A DONATION OF ONE PENNY FROM EVERY WORKING PERSON IN THE UNITED STATES WOULD MAKE YOU A MILLIONAIRE.

Out of a population of more than 300 million, there are only about 125 million people in United States who are actually working on any given day. If each of them were to send one penny to your bank account you would get 125 million pennies. That's $1.25 million. You'd be a millionaire.

Make those direct debits monthly, and you'd be a fat cat. Now *that*'s the kind of benefit I like.

24

HEAVENLY BODY

How much does the moon weigh?

The moon weighs about 65 billion billion tons. I know this because the width of my little finger at the third joint is about a third of an inch.

I once lived on a university campus in a row of terraced houses. We were bracketed by delightful neighbors who became lifelong friends. On one side was a neuroscientist who had voluntarily inverted his working day. Every evening, at about 9 p.m., I heard him start his car as he left for the quarter-mile journey to his research laboratory, and at 3 a.m. in the every morning I would be woken as he arrived back.

On the other side lived a mathematician. His route between home and office was determined by the principle that short cuts are a heresy against an ordered world. Each evening on his return from work, he walked home along the opposite side of the street. Arriving at a point precisely opposite his house, he would make a sharp 90-degree turn, and cross the road to his front door. Not for him the licentious joy of a walk along the hypotenuse.

Sandwiched between these archetypal academics, we lived very ordinary lives—apart from the grinding noise made by my telescope-making machine at the top of the stairs.

Some telescopes (and all large ones) use a convex mirror rather than a glass lens to gather light. They are called reflecting telescopes. Like a prism, a lens splits light into different colors. An image viewed through an inexpensive lens telescope—a refracting telescope—will show blurred colors on the edges of objects in the field of view. A mirror doesn't split light in this way, so the image in a reflecting telescope is free from unwanted colored fringes.

The simplest kind of reflecting telescope is called a Newtonian telescope. It has a concave mirror at the bottom end of the telescope tube. Light from whatever is being observed goes down the tube and is reflected back up to form an image somewhere just near the top end of the tube. A second mirror, small and flat and mounted at 45 degrees, interrupts the light path and directs the image out of the side of the tube where it is viewed through a magnifying lens—the eyepiece. A reflecting telescope is actually just a microscope—to look at the image—coupled with a large mirror to gather and concentrate the faint light from the night sky.

Of course, the surface of the curved mirror has to be highly polished. Most important, and something that takes a lot of care and effort to achieve, the profile of the curve must be highly accurate. But surprisingly, you can make a good telescope mirror in your own home, using your own hands and some basic equipment.

It's done by rubbing two circular disks of glass together in a sandwich filled with some sort of abrasive paste. The disks will wear into a convex mirror and a perfectly matching concave mirror. The mirror profile will be accurate to a few millionths of an inch, thanks to the unavoidable consequence of geometry, wear, and force between the two pieces of glass.

There is real excitement in looking at the moon and planets through a telescope, especially if you've made it yourself. Being outside on a cold night and knowing that the image projected on your retina is made by the impact of photons of sunlight that bounced off Saturn or Jupiter an hour or so earlier gives a sense of perspective on life.

The stars are all so distant that they remain points of light regardless of how much your telescope magnifies their image. By comparison, the moon is so close to the earth that a telescope isn't needed to see detail on its surface. Here's how to estimate the weight of the moon, using just your finger and no telescope.

Go outside on the night of a full moon and raise a finger in front of one eye until it covers the moon. Move your finger nearer or farther from your eye until it just covers the moon's diameter. At that point, make an estimate of the distance of your finger from your eye. Now go back inside.

The moon is about 250,000 miles from the earth. I found that my 1/3-inch-wide finger had to be held at arm's length to cover the moon. That's roughly three feet.

Now imagine two pieces of string (you'd need quite a few balls of it) attached to either side of the moon and stretched back taut to your eye. The strings meet at your eye, but three feet back toward the moon, they are already a third of an inch apart. After 9 feet they'll be 1 inch apart. Just work out how many steps of nine feet there are between the earth and the moon and you'll have the diameter of the moon in inches. There are 5,280 feet in a mile so the number of nine-foot steps is 250,000×5,280/9=146,670,000. The strings will then be separated by 146,670,000 inches. That makes the moon's diameter 2,310 miles. This estimation technique, based on *similar triangles,* can be applied to virtually any distant object as long as you know how far it is from your eye.

Back to the original question, which was to find the weight of the moon. We've estimated its diameter, and it's just a short step to get its weight. First, work out the volume in cubic yards, then estimate the weight of one cubic yard of

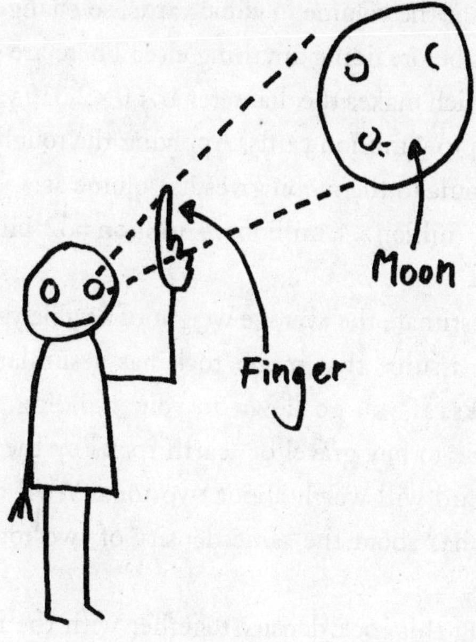

moon rock. Finally multiply the two numbers together to get the weight.

The volume of a sphere like the moon is one-sixth times π times its diameter, times its diameter, times its diameter. This formula is written in mathematical symbols as:

$$\text{Volume} = 1/6 \times \text{Diameter} \times \text{Diameter} \times \text{Diameter}$$

The mathematical constant π is about 3. We can take the $1/6 \times \pi$ in the formula to be 1/2, and the volume of a sphere to be roughly half the cube of its diameter. (Think about it: the volume of a sphere is about half the volume of a box into which it would just fit.)

We want the volume in cubic yards, so change the diameter to yards before doing anything else. There are 1,760 yards in a mile, which makes the diameter 1,760×2,310 yards. That's near enough to 4 million yards. Applying the rough-and-ready volume formula to the moon gives its volume as:

1/2×4 million×4 million×4 million=32 billion billion cubic yards

Now estimate the average weight of a cubic yard of moon rock. We'll assume that moon rock has a similar density to regular rocks. If you go down to your building supply yard you'll be able to buy gravel or "earth rock" by the cubic yard; one cubic yard will weigh about two tons. We'll assume that moon rock has about the same density of two tons per cubic yard.

Putting this rock density together with the moon's estimated volume of 32 billion billion cubic yards gives a weight of 64 billion billion tons. The presently accepted value is about 65 billion billion tons, so the estimate turns out to be quite good, considering it's based on the width of your finger.

SPEAK GEEK

— A practical application —

Sunlight is 400,000 times brighter than moonlight.

Light from the sun expands outwards into space. After 93 million miles, a tiny fraction lands on the moon, from which some of it is scattered back into space as moonlight. The total amount of scattered moonlight is determined by three factors: the area of the moon facing the sun, the moon's distance from the sun, and the reflectivity of the moon's surface. A small fraction of that moonlight lands on the earth's surface after a journey of around 250,000 miles from the moon.

The intensity of the sun's light that arrives on earth via reflection off the moon—moonlight—works out at about 2.5 millionths of the brightness of light coming directly from the sun itself. In other words, sunlight is 400,000 times as bright as its more romantic cousin, moonlight.

25

PASSING WATER

How many times has the River Thames been through the human body?

I was once taken by a student to his workplace: the sewage processing works on the edge of a provincial town. It was a sparkling, breezy autumn morning as we walked across one of the steel gantries spanning the filter beds. The morning was so sparkling and breezy that spray from nozzles on the rotating arms over the beds made a rainbow in the air above us. Drops of distinctly second-hand water blew onto my face.

As I tried not to open my mouth or lick my lips, the student ushered me to the place on the farm where the purified water was discharged into the local river. We leaned over the parapet and watched as electronic sensors continuously monitored the small but fast-flowing river generated by a town with a population of some 100,000 people. The sensors signaled to a computer that the effluent was now pure enough to drink. But I wasn't thirsty.

In London there's an urban myth that water in the Thames passes through a human body seven times before it leaves the city for the sea. The basis for this assertion is that water is extracted from the Thames for supply to Londoners and people living in the Thames Valley, who drink it, cook with it, and make tea with it. The water then returns via sewers to the river where it is extracted to start its next circular journey via the human bladder.

There may be something in the story. In 1852 an Act of Parliament was passed making it an offense to extract water from the tidal Thames in London. The act aimed to protect Londoners from disease, chemicals, and sewage carried by the river. As a result, two big water extraction plants were built. They still operate, upstream in the nontidal part of the Thames, isolated from London sewage by the last locks on the Thames, at the towns of Teddington and Richmond.

Getting enough clean water to people, even in rich and technologically developed countries, is difficult. It will get harder as populations increase, and harder still if the climate change projected by some scientists comes about, giving us hotter summers and uneven rainfall patterns. At best it will cause economic damage; at worst it will lead to war.

But how much of that dirty Thames water has really passed through human bodies? To get the answer, we need to know how much water passes through a single human in a day. We'll also need to be a bit more precise about who is consuming the water and which bit of the river we're talking about. A convenient boundary is Teddington, where the Thames becomes tidal. We'll do our calculation for the water and people upstream of this point. (A calculation on the river downstream of London wouldn't be very meaningful because the water flows in and out with the tide twice a day.)

The amount of effluent produced by each person in a day will be similar to the amount of fluid they take in by mouth. Of course, some will be lost as perspiration, but a large part will be disposed of in the toilet. You could look up the daily amount of water drunk or taken with food by a typical person, but it's easy to examine your own habits and arrive at the answer. I recorded my personal oral fluid consumption for a day and got this list:

TIME	DESCRIPTION	INTAKE METHOD	VOLUME
Breakfast	Tea/coffee	2 cups	0.50
	Juice	1 glass	0.25
	Milk	With cereal	0.25
Midmorning	Tea/coffee	2 cups	0.50
Lunch	Water/juice	1 cup	0.25
	Water	In cooked food	0.25
Afternoon	Tea/coffee	2 cups	0.50
Evening meal	Water	In cooked food	0.25
	Juice/milk/water	1 glass	0.25
Evening	Tea/coffee	2 cups	0.5
	Social beer/wine	2 glasses	0.5
Total water consumption			**4.0 liters**

Four liters passed through my body. You could live quite happily on four liters per day, but you actually have about 160 liters delivered to your house through the water company's pipes. It'll be similar in the United States. If water were delivered to your your door, there would be about 300 pint bottles, every day. The 156 liters of water that doesn't pass through you is used for flushing the lavatory, washing clothes and dishes, and a host of other essential and not-so-essential water purposes.

To find the total amount of water added to the river after passing through a human body, we need to know the number of people living in the Thames Valley, say upstream of Teddington Lock. It's safe to assume that the effluent from anyone living in the Thames Valley area ends up in the Thames—effluent will not flow uphill, out of the valley, of its own accord. A simple approach is to look at a relief map of Southern England and draw a line along the high ground to the north and south of the Thames, and identify the big centers of population lying between the lines.

We could take the population in the Thames Valley above Teddington to be roughly a million people. To get an exact value you'd have to use the relief map and make a detailed count of all the major towns lying in the valley. It could be 0.5 million and it could be 4 million, but we are just interested in getting an approximate answer, so a round figure of 1 million will do. The amount of water flowing daily into the Thames that has passed through a human body is therefore 1 million multiplied by 4 liters per person: 4 million liters.

It's more convenient to change that figure into liters per second. There are 3,600 seconds in one hour and 24 hours in a day, so in one day there are 3,600×24=86,400 seconds. To get the human water flow in liters per second, just divide the daily rate by this number. That's 4,000,000/86,400=46 liters per second. It sounds a lot, but how does it compare with the total flow in the nontidal part of the Thames?

If you have ever been out in a row boat on the Thames, you'll have found that there is a gentle current in the center of the river, of about 2 mph. That's about a meter per second. It's easy to confirm this by dropping a stick in the river and timing how long it takes to travel, say, 10 meters.

The river at Kingston is about 2 meters deep in the middle and about 50 meters across. If the depth were the same all the way across, the cross-sectional area of the river would be

$2 \times 50 = 100$ square meters. In reality, the river becomes shallower at the edges, making the cross-sectional area less—let's say 50 square meters.

The current is also much less near the bottom and the banks of the river than at the surface in the middle, so let's take the average current across the river to be half our initial estimate. That's 0.5 meters per second. On a rainy day the current will be much stronger because of all the water running off the surrounding land, but we'll assume that the river is in a hot, dry summer state.

If all the water in the river is flowing downstream at 0.5 meters per second, and the cross-section of the river is 50 square meters, then in one second the volume of water that passes a given point on the bank is $50 \times 0.5 = 25$ cubic meters. So, the flow rate of the nontidal Thames is about 25 cubic meters per second. One cubic meter of water is 1,000 liters, so another way of stating the flow rate is 25,000 liters per second. That is 6,000 gallons per second.

Our original question can now be answered. The average flow rate of water through the bodies of all humans living upstream of Teddington is about 50 liters per second, but the flow rate in the river is 25,000 liters per second. So only 1/500 of the water flowing past Kingston-upon-Thames has actually seen the inside of a human bladder within the last few days. Of course, over a period of millennia, it's likely that every drop of water in the river has seen the inside of someone's or something's bladder, but you probably won't be too worried by a biological contact that happened in the Jurassic Period.

But hang on—there's the other 156 liters of water used by each person every day. A significant amount of that is used to flush the lavatory. It's about 10 liters for a full flush, so you probably flush 50 liters of water down the potty each day. Then every time you have a bath you'll use another 50 liters. Your water consumption is already 100 liters a day, even without washing your clothes and dishes, not to mention the extravagances of car washing or watering the garden.

If we count the total volume of used water that's returned to the Thames, the myth looks a little bit closer to reality. One million people, each consuming 160 liters daily, gives a total of 160 million liters. That works out at around 2,000 liters per second, which is almost a tenth of the water flowing past Kingston-upon-Thames.

Remember this when you go swimming: 10 percent of the water around you has been through a lavatory, dishwasher, or washing machine. It is a remarkable feat of engineering and organization that for the past twenty years this water has been so well purified that it presents a negligible risk to your health.

SPEAK GEEK

— *A practical application* —

A THUNDERCLOUD CAN HOLD MORE THAN 1,000 TONS OF WATER.

All the water we will ever consume has spent time as rain clouds. But how much water can a cloud hold? A true Geek will have a go at working it out, but will also recognize that the estimate could easily be ten times too big or ten times too small.

Cloud thickness ranges from hundreds to thousands of feet; for simplicity, let's go for a 1,000-foot (1/5 mile) thick, 1 mile long and 1 mile wide cumulus cloud. Multiply them all together and you've got a total volume 1/5 cubic miles. That's around 30 billion cubic feet.

So how much does a cubic foot of cloud water droplets weigh? The other day, I heated some water inside a freezing cold shed. After about 1/10 of a pint had boiled away, inside the shed was much like being on a mountaintop in thick mist. The volume of my garden shed is 1,500 cubic feet, so the density of water droplets was around 1/15,000 pints per cubic foot. So, maybe the cloud contains 30 billion × 1/15,000 = 2 million pints of water. That's 1,000 tons to you and me.

26

THE MAN IN WHITE

How much could sea levels rise?

In 1959 I was lying in a bed in Great Ormond Street Hospital for Children. It was early evening. Matron appeared, accompanied by a nurse. She checked that the area around my bed was clean, briskly straightened the bedclothes, and then, with a stern face, brushed my hair. Curtains were pulled quickly around the bed and she was gone. The nurse stayed to whisper, comfortingly, that the *Consultant* was coming.

The curtains were pulled aside, and there he was, surrounded by a group of respectful trainee doctors, each in a white coat and wearing their stethoscopes round their necks. The Consultant was the Big Man, and he was the only person not wearing a white coat. Even to an eight-year-old, he was beyond question The Man Who Knew Everything. His words were The Truth.

That encapsulates how we handle complex bodies of knowledge and skill. We hand over the responsibility for knowing and doing to trusted individuals or groups, just as priests were once believed to hold special knowledge and were therefore entrusted with making our connections with gods and the elements.

Often, those trusted individuals have special regalia to mark them out—the equivalent of a priest's robes. Doctors and scientists wear white coats, the academic professor wears a bow tie, the judge wears a black robe. Even a fluorescent orange safety jacket is a signifier of authority.

But handing over responsibility for identifying truth to doctors, scientists, journalists, or whomever is dangerous unless you are able to make basic logical checks on their pronouncements. If you don't already doubt professional wisdom, just think about the number of changes there have been over the years in professional advice on which way up to place your baby in its crib or the safety of lead additives in gasoline, or the scientifically flawed study on supposed links between autism and mumps, rubella, and measles vaccination. A British study claimed to show links between vaccination and autism. It was wrong.

In medicine, the kind of sanity checks we can all carry out without possessing expert knowledge are usually simple statistical questions. How many people took part in the trial of a new therapy? How were they selected? Is the reported effect causal or just associative?

In the environmental sciences, the tests involve rudimentary, commonsense physics. Anyone can check the logic of expert pronouncements. For example, take this imaginary newspaper headline:

ARCTIC SEA ICE MELTS: MASSIVE FLOODING THREATENED AS SEA LEVEL RISES

It's obviously true: everyone knows that ice is made of water, and if melted, that water is added to the ocean and will raise the sea level. But what kind of ice are we talking about?

Most of the Arctic ice visible from space is sea ice. It is floating in the water of the Arctic Ocean, and if floating sea ice melts it will make absolutely no difference to the volume of the ocean, or to its level. The Greek philosopher Archimedes pinned this down in what we now call Archimedes' principle.

The gist of his famous Eureka! moment is this. Anything that floats—a ship, a twig, a swimmer—displaces a volume of water the weight of which is exactly equal to the weight of the floating object. A floating iceberg is already displacing water with its weight. Melting the iceberg doesn't change the weight of water that it originally contained, so the sea level won't change.

If you want to test Archimedes, do an experiment in your kitchen. Put half a dozen ice cubes into a plastic measuring jug. Fill up the jug with water to, say, the 500 ml mark (about two cups), and wait. The ice cubes will float in the water; with about 10 percent of them above the water and 90 percent below, just like a real iceberg. When they have completely melted, check the level of water in the jug. If it has risen, you will have confounded a two-thousand-year-old scientific principle. If it hasn't, you will have proved that melting sea ice will not cause flooding.

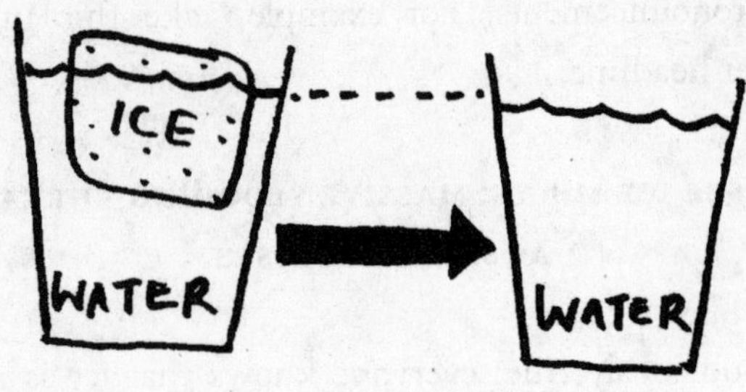

Here's a similar sounding but physically very different headline:

ANTARCTIC ICE MELTS: MASSIVE FLOODING THREATENED AS SEA LEVEL RISES

Now, unlike polar ice at the North Pole, the ice at the South Pole is resting on land: on the continent of Antarctica. If this ice melts, it really will add water to the oceans. But how much will the sea level rise as a consequence?

We could take a worst-case scenario and work out how much the sea level would rise if all the Antarctic ice melted, however long that took. To work this out we need two pieces of information: the volume of water in the ice resting on the continent, and the area of the world's oceans. Dividing the volume of melted ice by the ocean area will give the approximate rise in sea level. (The actual rise will be slightly less than calculated because land will be flooded as the sea rises, and a very small amount of the water from the melted ice will be used to fill this new bit of ocean volume.)

To picture how the method works, imagine a garden pond whose area is 100 square feet. Imagine also that you have a cube of ice with a volume 1 cubic foot. First melt the block of ice. It will shrink a bit as it melts because ice takes up more room than water, and you will end up with about 0.9 cubic feet of water. Now pour the water into the pond. The volume of the added water will raise the water level by 0.9/100=0.009 feet; about 1/10 of an inch.

The method seems to work, so let's apply it to the oceans. First, what is the volume of ice on the Antarctic continent? That's worked out by multiplying the area of the continent by the average ice thickness. You could look up these figures; the area is about 5 million square miles, and the average ice thickness is 1 mile. That gives a total ice volume of 5 million cubic miles of ice. The ice will shrink as it melts, and the water volume will be about 90 percent of the ice volume; 90 percent of 5 million cubic miles of ice is, still near enough, 5 million cubic miles.

Now we need the surface area of all the earth's oceans, to which the melted ice water will be added. About 70 percent of earth is covered by water, and the total surface area of the oceans is around 140 million square miles. Dividing the melted ice volume of 5 million cubic miles by 140 million square miles tells us how much the sea level would rise if all the Antarctic ice melted. It's 5/140=1/28 miles.

That's 189 feet. So, if all the Antarctic ice melted, much of London would be submerged. This is a pessimistic calculation. It's very unlikely that all that ice is going to melt any time soon. There is even evidence that the ice cover in parts of Antarctica is on the increase. But we do seem to be in a period of global warming, which could cause sea levels to rise, and this can happen without any ice melting anywhere. Imagine another news headline:

OCEANS HEAT UP: MASSIVE FLOODING THREATENED AS SEA LEVEL RISES

Water expands as it warms up. The volume of the water in the oceans will increase as it is warmed, and that will make the sea level rise. This will happen regardless of any increase in volume caused by ice melting.

Although it's one of the most familiar substances of all, water is quite strange in several respects. One of its peculiarities is the way in which its density changes with temperature. At just above the freezing point, one kilogram of water has a volume of 1.0002 liters. Increase the temperature to 4 degrees Celsius, and the same kilogram shrinks to almost exactly 1 liter. Above 4 degrees its volume starts to increase again, by roughly 0.04 percent for each degree of temperature rise.

The figure of 0.04 percent volume rise per degree gives us a way of working out how much the sea level would rise as the water expanded. We need to know the total volume of earth's oceans, the probable rise in temperature of the water due to global warming, and the oceans' total area.

The total area of the oceans is 140 million square miles. Their combined volume is the average depth multiplied by the area. Living in Britain on an island on the continental shelf, I am used to sea depths of less than 600 feet, but the average depth of the Atlantic is over 2 miles. The Pacific is even deeper: 2.7 miles, on average. The Indian and Southern Oceans are similar. In fact, the deepest point of the Pacific Ocean is almost 7 miles, about the same distance below the surface as jet airliners fly above it. If you sat at the bottom of this ocean trench and looked up, a jumbo jet floating on the surface (if you could see it, which you actually couldn't) would look the same size as those pinpricks of silver you see in the sky.

A fair estimate of the average ocean depth over the whole globe is 2.5 miles. With an area of 160 million square miles, the volume works out at 2.5×160 million$=$350 million cubic miles. Raise the temperature of the whole ocean by 1 degree, and its volume will increase by 0.04 percent. That's an increase of 350 million$\times 0.0004 = 140{,}000$ cubic miles.

The increase in volume will be spread over the entire area of the earth's oceans. The resulting sea-level rise caused by a 1 degree increase in global temperature will be 160,000/160,000,000 miles, or 5 feet. According to the gloomiest predictions, the global temperature will rise by about 4 degrees over the next century, in which case we could expect the seas to rise by $4 \times 5 = 20$ feet, simply as a result of their expansion.

But how soon is this going to happen? Should you think about moving to a higher area before house prices rocket? Here's another headline:

EARTH IN MELTDOWN: ANTARCTIC ICE WILL DISAPPEAR IN FIFTY YEARS

The rate and causes of global warming are still fiercely disputed. Most climate models point to increased atmospheric carbon dioxide, possibly coupled with a natural periodic increase in the sun's power as a cause of global warming. The mechanics of our biosphere—the atmosphere, ocean, surface environment, and life viewed as a single system—are so complex that no simple back-of-the-envelope calculation will tell us whether the current predictions of climate scientists are correct.

Still, you can attempt to test that last headline about all the Antarctic ice melting within fifty years. An important and obvious cause of ice melting is direct heat from the sun. As the ice melts, particles of dirt that were trapped inside the ice start to build up on the surface, which thus grows darker. Dark rocks are exposed at the edges of the melting ice sheet. The darkening surface absorbs more of the sun's heat and reflects less back into space. This further hastens the process of melting, leading to an unstable positive feedback effect.

Some climate scientists believe that this is what they call

a tipping-point mechanism of climate change that might lead to irreversible melting of the ice sheet. There are processes other than direct solar heating that might cause the ice to melt, for example hot winds from temperate regions, or simple proximity to an Antarctic atmosphere whose temperature has been raised above freezing by global warming. New snowfalls would tend to cover up the exposed dark surface of the melting ice to make it more reflective and counter the melting, but overall, the melting processes might be overwhelming.

Can we work out how much heat energy would be needed to melt the entire Antarctic ice sheet? We have already found that its volume is about 5 million cubic miles, and that's a weight of about 25 million billion tons. If the amount of heat needed to melt 1 ton could be calculated, we'd be able to work out the energy needed to melt the lot.

One pound of ice needs about 150,000 joules of heat to change it from ice to water. That's the heat you would get by running a 3 kW electric heater for about 50 seconds.

You can do a simple check on this figure next time it snows: empty an electric kettle and then stuff it full of snow. Switch on and time how long it takes until all the snow has just melted. Then weigh the water. For every pound weight of water, the kettle should have taken about 50 seconds to do the melting. To melt 1 ton of snow, your kettle would be going for 100,000 seconds—28 hours.

If—and it's a big if—the energy necessary to melt the ice in Antarctica arrives mainly as heat from the sun, we need to know how much heat energy the sun delivers to the earth's surface. We can start at London on June 21, the Summer Solstice. At noon, if the air is very clear, the sun will throw around

1,500 watts of heat onto each square yard of the ground. That's about half the heating effect you would get from standing in front of an old-fashioned radiant electric heater. On a cloudy day in January it's only about 20 watts per square yard.

At night, of course, there is no incoming heat. In fact, the process is reversed: if the air is clear, heat is radiated back into space during the night. That's why there is sometimes a ground frost, even when the air temperature is above freezing. Overall, it is estimated that the average power of the solar heating in London is 100 watts per square yard. That's averaged over twenty-four hours a day, all year round.

At midsummer in London the sun reaches an altitude of about 61 degrees, but in Antarctica it is much lower in the sky, and its power is much less. In fact, from the South Pole the sun's altitude is only 22.5 degrees at noon on June 21. That's only a little more than in London at noon in midwinter.

Based solely on the angle of the sun and the lengths of the days throughout the year relative to London, the average solar heating at the South Pole will be about 40 percent of the London value: about 40 watts on every square yard. There are other factors to take into account, such as the greater thickness of the atmosphere through which the low-altitude sun's rays have to travel in Antarctica. But the atmosphere may be much clearer in Antarctica than over London, so we'll keep 40 watts per square yard as our final figure.

Clean snow can reflect about 90 percent of the incident rays, and even dirty snow and ice will reflect some of the heat. The exact amount will depend on how dark the ice and snow become. We'll assume a figure of 75 percent, although actual measurements are needed to find the true value. That would mean that 25 percent of the sun's heat is absorbed by the dirty ice and causes melting, which gives us about 10 watts per square yard.

We have already calculated that it takes a 3,000-watt heater about 50 seconds to melt a pound of ice. The Antarctic sunlight's 10 watts per square yard of absorbed power will take $3{,}000/10\times 50=1{,}500$ seconds to melt 1 pound. Very roughly, that's 1 pound melted every half hour in each square yard of Antarctica.

A pound of ice has a volume of 1 pint, and when it's melted it will form a layer of water 1/50 of an inch thick over the 1 square yard area. So, it looks as if the sun might melt 1/25 inch of ice every hour. With around 9,000 hours in a year, the yearly melt depth would be 360 inches, or 30 feet. It sounds a lot, but the average ice sheet thickness in Antarctica is 1 mile, and, with our assumptions, it would take 176 years to melt completely.

So, our imaginary headline

EARTH IN MELTDOWN: ANTARCTIC ICE WILL DISAPPEAR IN FIFTY YEARS

is almost certainly wrong.

In fact, this last simplistic estimate shows that sometimes experts are just what we need: more sophisticated predictions by climate scientists and geophysicists put the time to melt the ice sheet at nearer a thousand years. That's still potentially very serious, but not imminent.

Thank goodness for experts—sometimes.

SPEAK GEEK

— *A practical application* —

TWENTY NEW TREES NEED TO BE PLANTED EACH YEAR TO OFFSET YOUR CO_2 EMISSIONS.

Wood is composed mainly of a substance called lignin cellulose. A fully grown pine tree contains up to half a ton of the stuff. About 44 percent of the weight of each molecule of cellulose is made of carbon, so the tree contains about a quarter of a ton of carbon. That quarter ton will have come from just under 1 ton of atmospheric CO_2 while the tree was growing.

The average American is responsible for 20 tons of CO_2 emission each year. In that case, you'll need to grow twenty new trees each year to be carbon neutral. This would mean planting and maintaining 6 billion new trees every year. The spacing between mature pines in a forest is 4 to 5 yards, so each tree occupies a ground area of around 20 square yards. In round figures, the area of land needed for new forest will be 180 billion square yards. That's one new forest 240 miles long and 260 miles wide planted every year.

Texas would be completely covered in forest in less than five years. You wouldn't see the wood for the trees.

Acknowledgments

This book wouldn't have been started or finished without the encouragement and support of my wife, Seija.

The seed of the idea wouldn't have been presented to the publishers if my agent, Cathryn Summerhayes, hadn't recognized that this was a book worth writing and helped enormously in the development of the initial drafts.

And, once the writing started in earnest, my editors, Ben Dunn and Jack Fogg, provided constant and invaluable guidance about what works and what doesn't, as well as helping make dry calculations into something more fun.

My friends Mike Carey, Ramsay Gibb, David Vincent, and Richard Aldridge all helped me with their own personal contributions, and John Woodruff provided the final painstaking and tolerant editing.